BUILD A WORKING MAGLEV TRAIN

MAGNETIC SCIENCE

PLUS 8 OTHER MAGNETIC EXPERIMENTS

MAGNETIC SCIENCE

AUTHOR NANCY DICKMANN

ILLUSTRATOR MICHAEL DRIVER

Quarto is the authority on a wide range of topics.
Quarto educates, entertains and enriches the lives of our readers—enthusiasts and lovers of hands-on living.
www.quartoknows.com

Author: Nancy Dickmann
Illustrator: Michael Driver
Consultant: Dr Charlotte Sleigh
Design and Paper Engineering: Alan Brown
Editor: Harriet Stone
Publisher: Jon Gilbert

Part of the Quarto Group
The Old Brewery, 6 Blundell Street,
London, N7 9BH

First published in the United States in 2021
by Quarto Children's Books, an imprint of The Quarto Group,
26391 Crown Valley Parkway, Suite 220
Mission Viejo, CA 92691, USA

A CIP record for this book is available from the Library of Congress.

ISBN: 978-0-7112-8032-8

First printing, June 2022
Printed in Shenzhen, China HH062022

9 8 7 6 5 4 3 2 1

CONTENTS

MAGNETS AND MAGNETISM

Long ago in northern Greece, a man named Magnes was keeping watch over his flock of sheep. He realized that he couldn't move—the nails in his shoes and the metal tip of his shepherd's staff were stuck fast to the black rock that he was standing on. It was a mineral called magnetite, which can sometimes form a lodestone, a type of natural magnet.

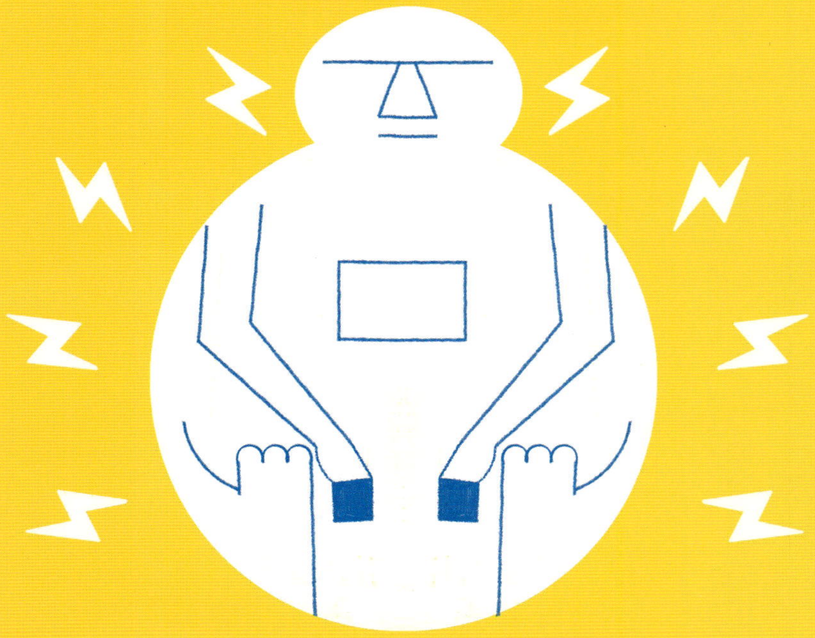

The "fat boy" statues of Guatemala are made from lodestone, proving that our ancestors knew about magnetic properties 4,000 years ago!

MAGIC OR SCIENCE?

The tale of Magnes is a great story, but it's probably not true. What is true, however, is that the ancient Greeks were aware of the power of lodestones to attract some types of metal. They had no idea how it worked, though. About 2,600 years ago, a philosopher named Thales of Miletus theorized that lodestones were able to attract iron because they had a soul.

PUTTING MAGNETS TO USE

As the centuries passed, people learned how to use the power of magnets. The ancient Chinese discovered that a thin piece of lodestone placed in a bowl of water would point north and south, and a lodestone fashioned into a ladle shape would spin on a plate to point north. This could help sailors to find their way. But it was about 1600AD before scientists began to really understand how magnets worked.

The ancient Chinese used compasses such as this one to help with fortune-telling.

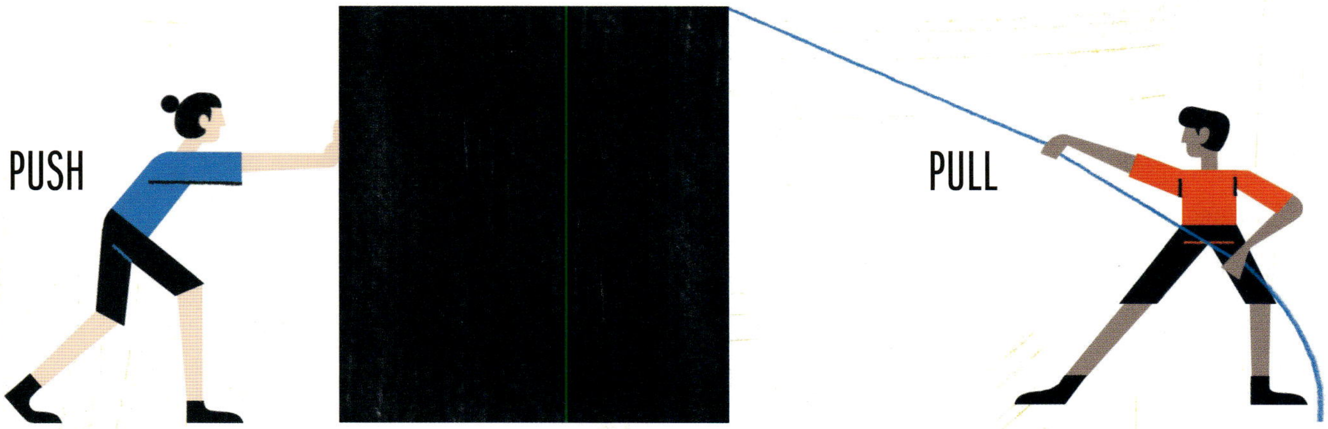

A FORCE OF NATURE

So what exactly makes magnets work? Despite what the ancients thought, this question has a scientific answer. Magnetism is a natural force that can be felt by some types of metals. It is one of the fundamental forces of nature. Forces are often defined as pushes or pulls, and a magnet can do either. It can pull some materials toward itself, and push others away. And it can do this without even touching them!

MAGNETIC MACHINES

Magnets are found inside all kinds of machines around the home. Household objects that contain magnets include vacuum cleaners, washing machines, speakers, fans, refrigerators, computers, and cell phones. Where would we be without them?

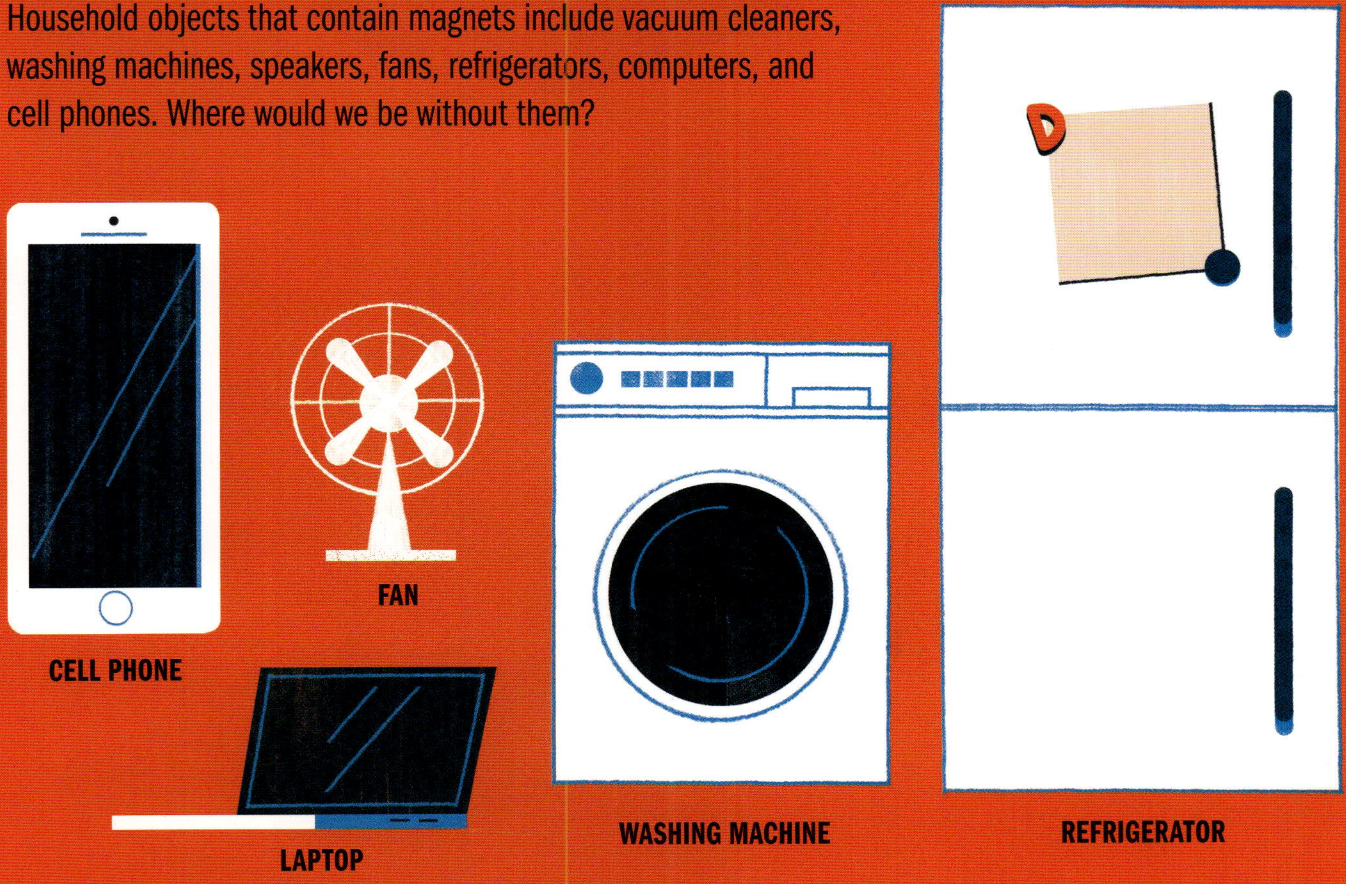

ELECTROMAGNETISM

Magnetism is actually part of a larger force called electromagnetism. It's sort of a combination of electricity and magnetism. In the 1800s, a Scottish scientist named James Clerk Maxwell discovered that the two phenomena are actually different sides of the same effect.
Learn more about electromagnets on page 32.

PULLING POWER

TESTING MATERIALS

A strong magnet can pull on an object through another material, and even pull on things without touching them. This may be why people in ancient times thought they were magic! Test out different materials in this pulling power experiment.

YOU WILL NEED:

- 1 disc magnet
- several paperclips
- cardstock piece 1 from the kit
- glass tumbler
- materials to test, such as fabric or plastic

1 Hold the magnet about an inch above a paperclip. How close does it have to be before it will rise up to stick to the magnet?

2 Put the disc magnet on top of piece 1, then hold paperclips underneath. When you let go, they should stay where they are. Can you make the paperclips move by pulling the magnet around the cardboard?

3 Put a few paperclips into the glass tumbler and see if you can use the magnet on the outside of the glass to pull them to the top.

4 Test out other materials, such as a cotton towel or a piece of thin plastic.

HINT

If the magnetic field is strong enough, a magnet can attract things through different materials. Is there any material that the magnet doesn't pull through? Does the thickness of the material make a difference?

LEVITATING ROCKET

Make a rocket float in the air with this amazing levitation experiment.

YOU WILL NEED:

- 2 disc magnets • string
- cardstock pieces 2–9 from the kit
- 1 paperclip • tape

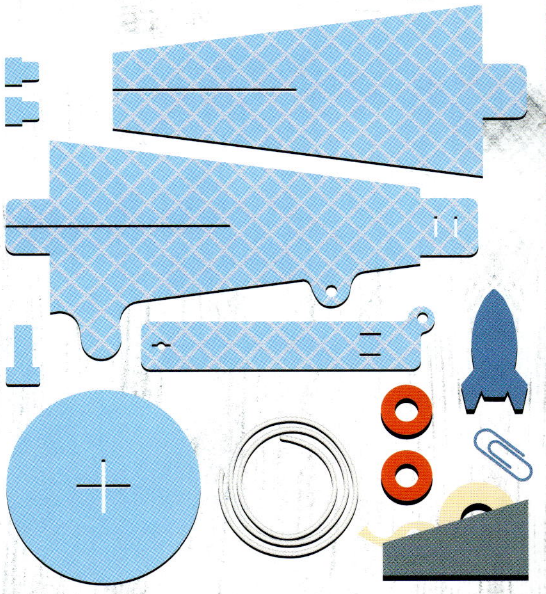

1 Use cardstock piece 2 as a template and cut out a rocket from paper. Tie a piece of string to the bottom of the paperclip and then attach it to the very top of the paper rocket.

2 Build the crane arm using pieces 3 to 8 from the kit (see page 22).

3 Slot the end of the string through the hole in the tray and feed it through so that it's hidden underneath.

4 Slot piece 9 through the end of the crane arm. Attach the two disc magnets using tape.

5 Carefully hold the rocket up close to the disc magnets, until the string is taut. Let go and your rocket should "levitate" in the air!

HINT

A magnetic field can work from a distance and overcome gravity. The paperclip is attracted by the magnet's magnetic field, but the string keeps the paperclip from being pulled all the way up to the magnet, so the rocket seems to levitate in the air.

HOW DO MAGNETS WORK?

In the pulling power experiment you just tried, the paperclips stuck to the magnet. But other materials, such as cardboard, glass, and fabric, don't. Why is it that some materials are attracted to magnets and others aren't? And what makes a magnet magnetic in the first place?

MAGNETIC METALS

Only a few types of metals are attracted to magnets. No other substances are. Iron is attracted to magnets, and so is steel, which is mainly made up of iron. (Most paperclips are made from steel.) Nickel and cobalt are also attracted to magnets. Other metals, such as aluminum, are not.

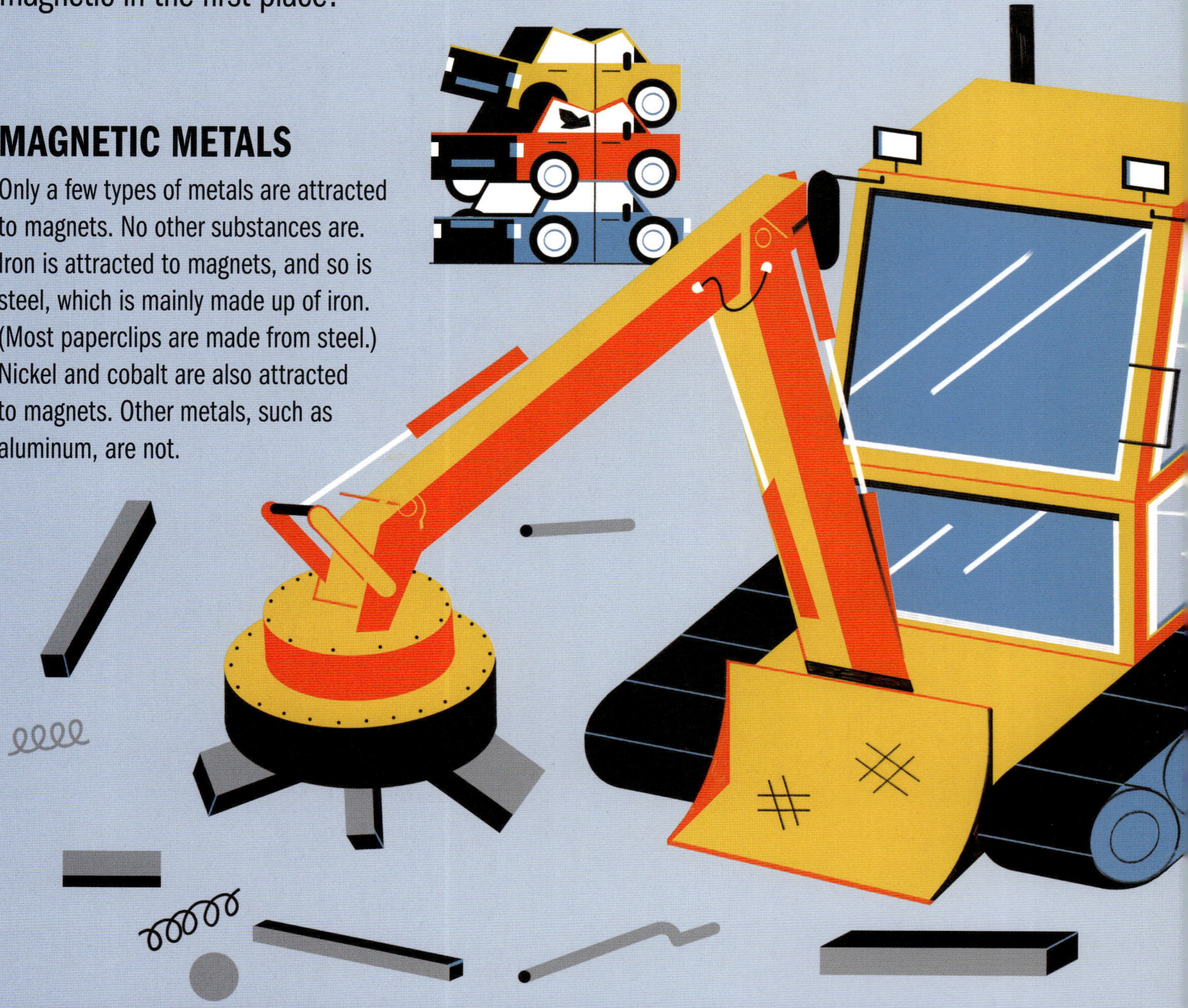

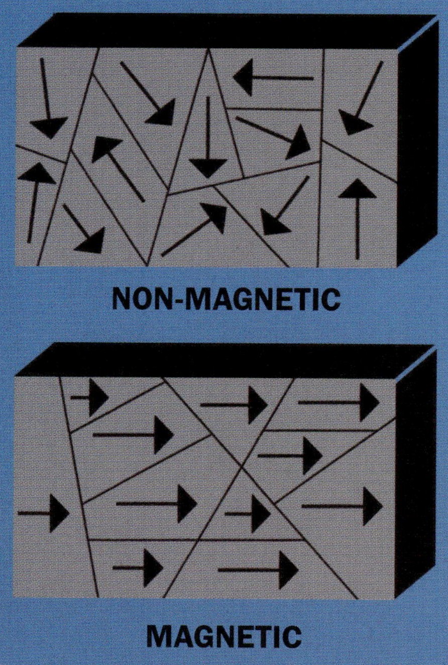

INSIDE A MAGNET

To understand what makes a magnet magnetic, you have to look really deep inside, to parts of the material that are too small to see, even with the most powerful microscope. Any material is made up of tiny regions called domains. Each domain has its own tiny magnetic field. In most substances, the domains are jumbled up, with the magnetic fields pointing every which way. This means the magnetic fields cancel each other out. But in a magnet, the domains are all aligned in the same direction. Their individual magnetic fields combine to make a magnetic field that is big enough to have effects.

TEMPORARY AND PERMANENT

Some magnets are always magnetic. They are called permanent magnets. Other materials can become temporary magnets. For example, if you pick up a paperclip with a magnet, you can usually get it to pick up a second paperclip. The first paperclip has become a temporary magnet. But once it is no longer stuck to the permanent magnet, it will no longer be magnetic. Electromagnets are another type of temporary magnet. They only remain magnetic as long as the electric current is switched on.

MAGNET NO MORE

Many permanent magnets gradually lose their strength over time. It is also sometimes possible to make them lose their magnetism more quickly. Heating a magnet, repeatedly hammering it, or dropping it onto a hard surface may demagnetize it.

MAGNETIC FIELDS

The cool thing about magnets is that their power of attraction can work across empty space, as we saw in the rocket experiment on page 7. A magnetic field is a type of force field produced by a magnet. You can't see, smell, or touch one, but you can easily see its effects.

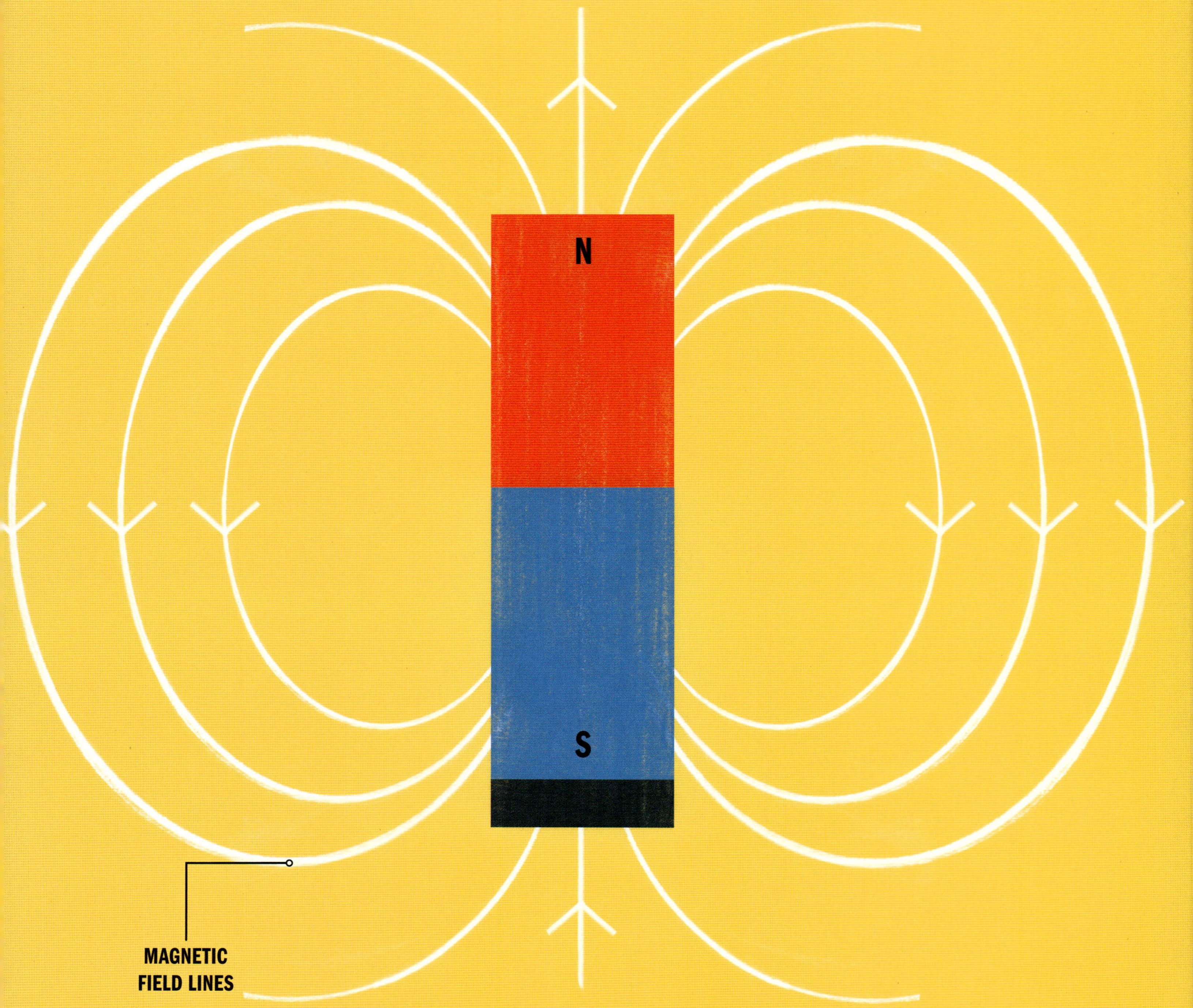

POLE TO POLE

All magnets have two poles: north and south. In a bar-shaped magnet, one end is the north pole and the other is the south. The invisible lines of a magnetic field point away from the north pole of a magnet and toward its south pole. If you scatter iron filings on a flat surface and then place a magnet among them, the filings will move. They will arrange themselves in a pattern that follows the invisible lines of the magnetic field. See this at work on page 12.

ONE BECOMES TWO

If you break a bar magnet in half, you won't get two magnets that each have a single pole. Instead, each of the two pieces will have a north and south pole.

ATTRACT AND REPEL

Have you ever tried to push two magnets together and discovered that they won't touch? It feels like they are pushing back. A magnetic field exerts a force that can be either a push or a pull. The north pole of a magnet will repel, or push away, the north pole of another magnet. A south pole will repel another south pole. (This is how maglev trains are lifted off the track.) But opposites attract—a north pole will attract the south pole of another magnet.

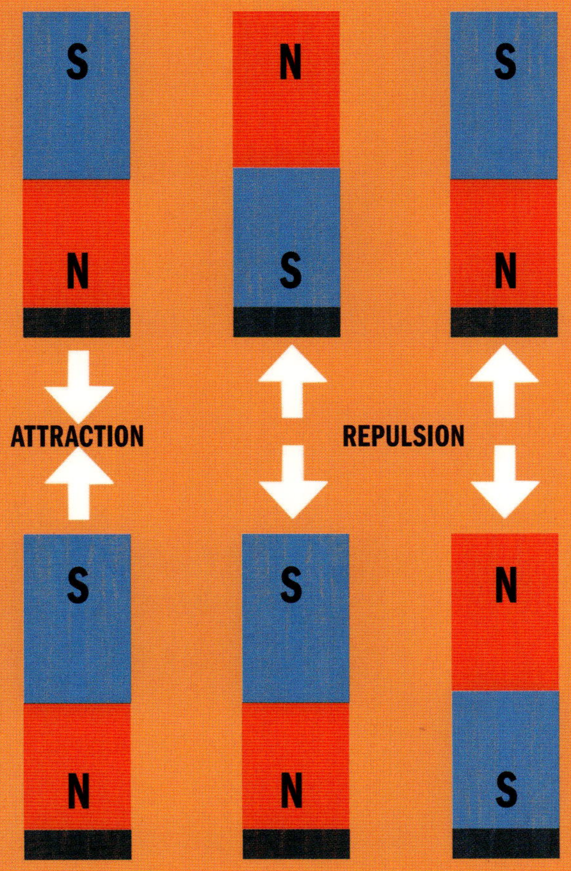

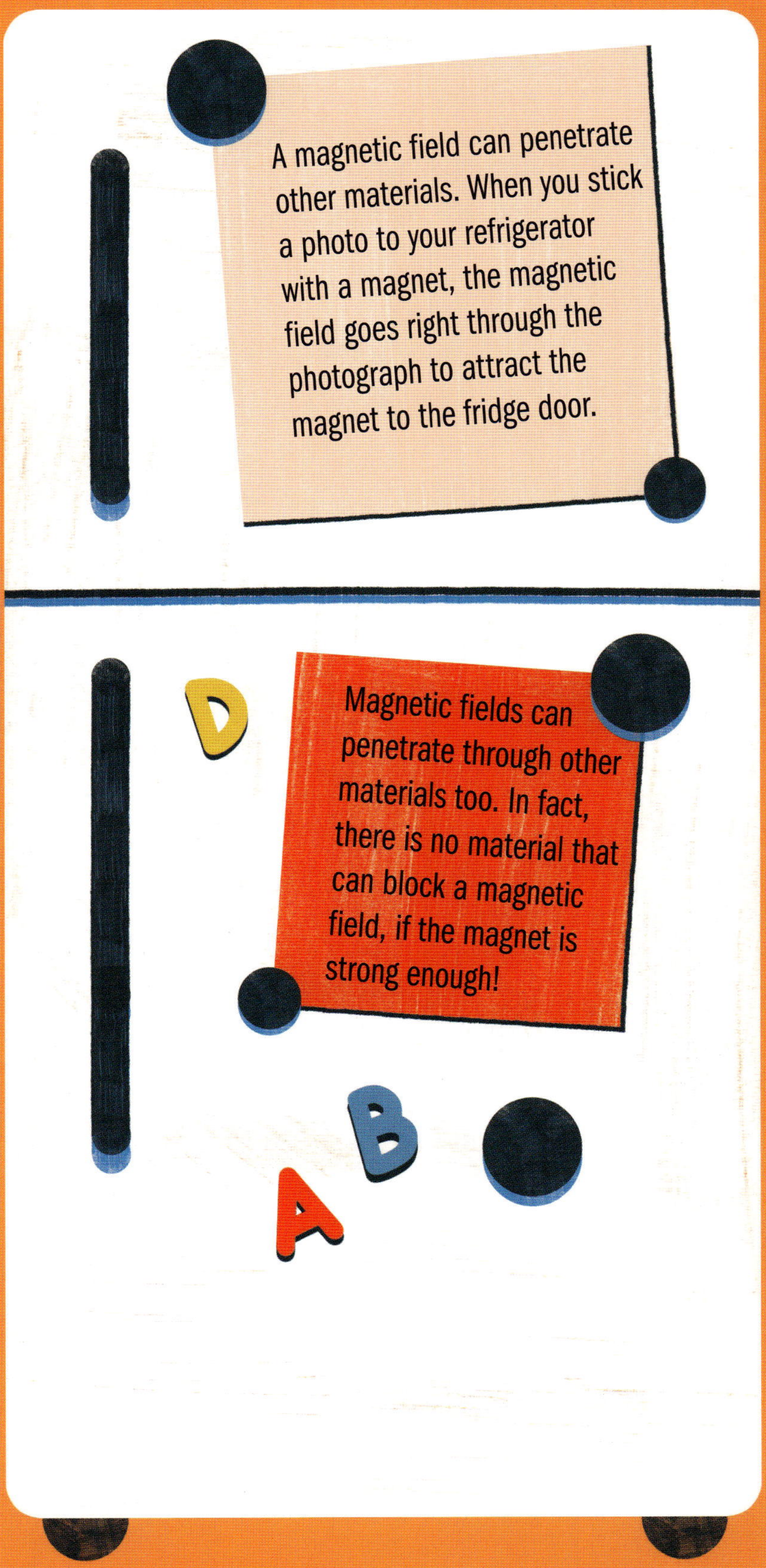

HOW METAL DETECTORS WORK

Magnetic fields help metal detectors find buried treasure. A wire coil in the detector creates a magnetic field. When it passes over a metal object, this makes an electric current flow through it, creating another magnetic field. A second coil in the metal detector is able to detect this magnetic field, and the machine beeps.

MAGNETISM AND PLANET EARTH

Earth is pretty amazing. It has towering mountains, vast oceans, and is teeming with plants and animals of all kinds. But our home planet is also hiding a truly awesome secret. It's actually a giant magnet!

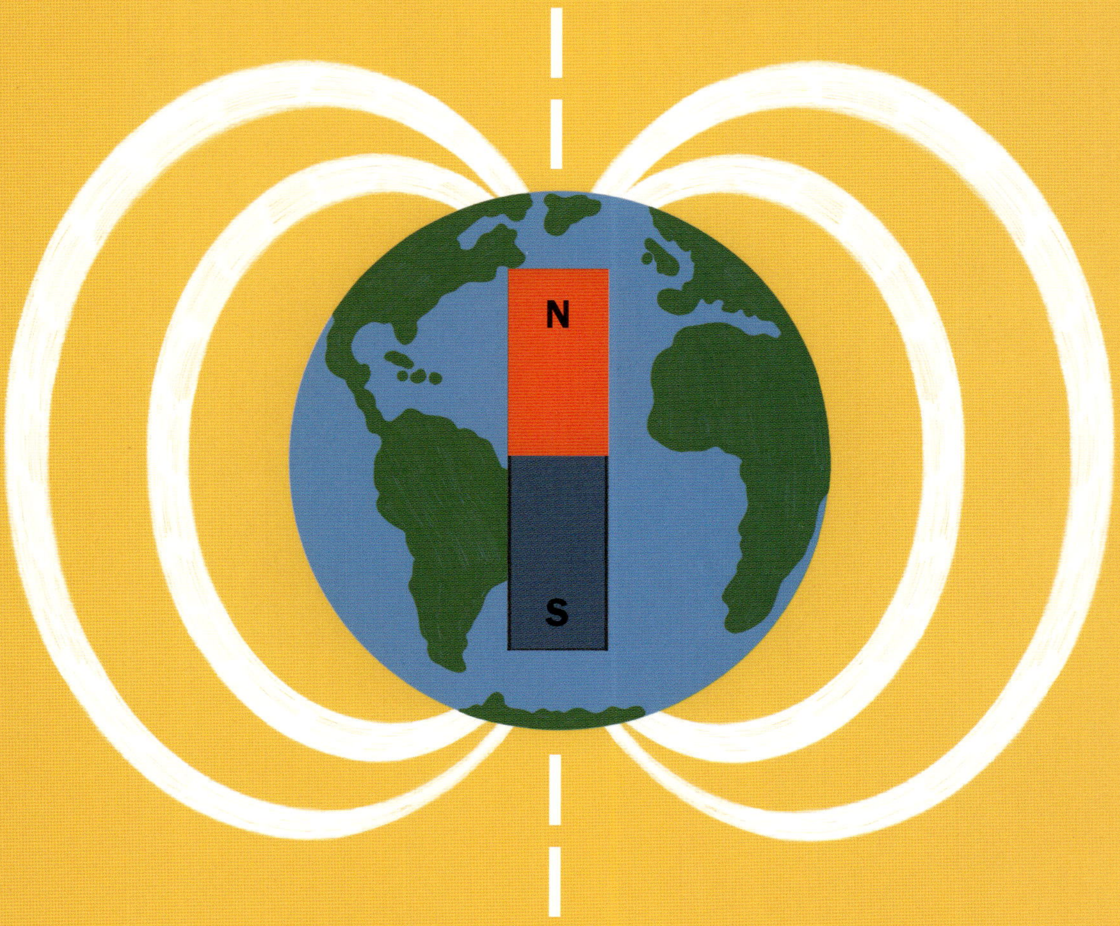

GOING THROUGH

You've probably heard of Earth's north pole and south pole—they are the points at the very top and bottom of the planet. Earth rotates (spins) around an imaginary stick called an axis that goes through the poles. Did you spot the clue in their name? Earth is a magnet with a north and south pole, just like the bar magnets you used in your experiment. It produces a huge magnetic field that surrounds the entire planet.

AURORAS

In Earth's far north and far south, at night you sometimes see curtains of shimmering greenish light dancing across the skies. These are the auroras, often known as the "northern lights" or "southern lights." They are caused by high-energy particles from the Sun interacting with Earth's magnetic field.

JOURNEY TO THE CENTER OF THE EARTH

The secret behind Earth's magnetism is hidden deep underground. Earth is made up of different layers. We live on the hard outer shell, which is called the crust, which sits on top of a layer of hot rock called the mantle. At the center is the core—a solid, super-hot ball of iron and nickel. It is surrounded by a layer of the same materials, but in liquid form. These liquids slosh around as Earth rotates, creating an electric current that produces a magnetic field.

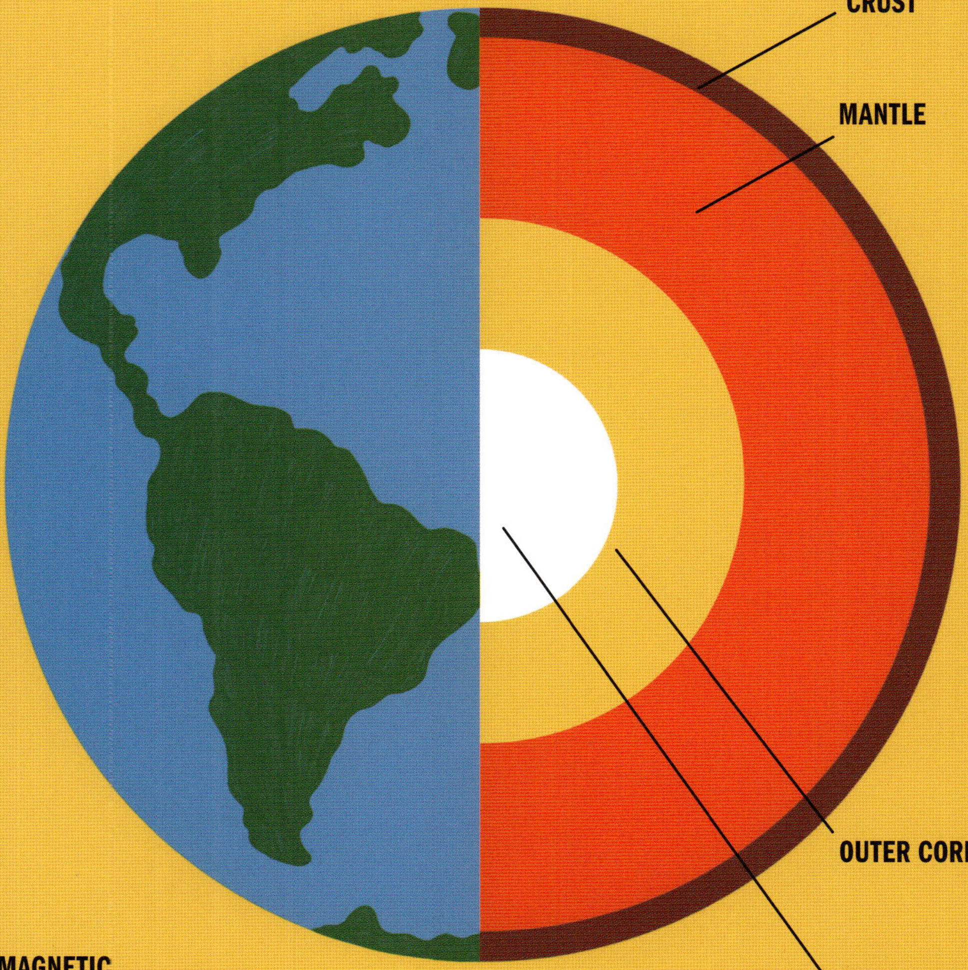

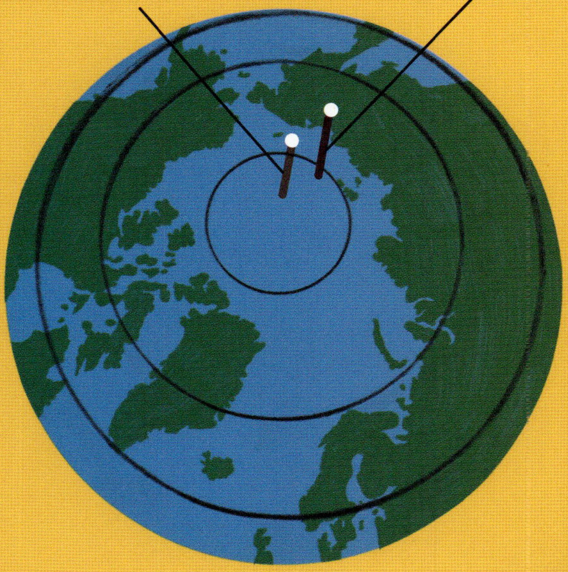

SHIFTING POLES

The geographic north and south poles are always in the same place. But Earth's magnetic poles move around, by an average of about 25 miles (40 km) per year. They are usually fairly close to the geographic poles, but not always. The swirling of the liquid core causes them to move.

ALL CHANGE

Geologists have discovered that Earth's magnetic poles can sometimes swap. The north magnetic pole becomes the south, and vice versa. This has happened at least 400 times so far, and the last swap was 780,000 years ago. We may be due for another swap in the next few thousand years.

POINTING THE WAY

For centuries, sailors and other travelers have used compasses to show them the way. These tools have allowed people to explore new places and map them. But they wouldn't work at all if it weren't for Earth's natural magnetism.

Lodestone floating in water

THE FIRST COMPASSES

We don't know for certain who invented the compass. Historical records show that people were using them to find their way in China and Europe at some point before 1200AD, though some historians think that the ancient Chinese used them much earlier than that. Early compasses used a piece of lodestone floating in water, which would point north. People later learned to use a lodestone to magnetize an iron needle as a pointer.

HOW DOES A COMPASS WORK?

A compass is a very simple instrument. All you need is a magnetized needle or pointer that is mounted on something so that it can swing freely. The compass often has a card with the different directions marked, to make it easier to read. Many modern compasses have the magnet inside a marked sphere that spins around within a liquid-filled capsule.

USING A COMPASS

To find your way using a compass, hold it flat and wait until the needle stops moving. The end of it that is painted red is now pointing north. Then carefully rotate the whole compass—the needle should keep pointing north while the rest of the compass moves. When the line marking north on the compass aligns with the needle, you can read the other directions: east, south, and west.

NORTH OR SOUTH?

It seems crazy, but Earth's north pole is actually a south magnetic pole! Opposite poles attract, remember? The north pole of the magnet in the needle of a compass is attracted to the south pole of Earth's magnet. So, when it points north, it is (sort of) pointing south!

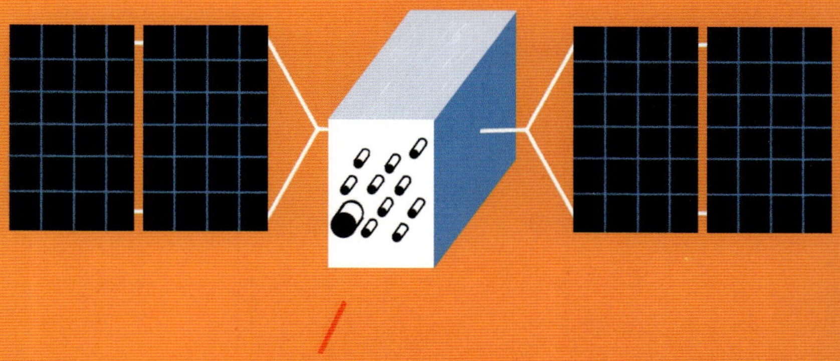

COMPASS VS. GPS

These days many people use GPS instead of a compass to find their way around. This system is not based on a compass reading. Instead, it uses a network of satellites in orbit around Earth. A GPS receiver can take signals from three satellites to pinpoint your location.

MAKE A COMPASS

You'll be ready to go exploring once you've followed these instructions to make your own magnetic compass!

YOU WILL NEED:

- 2 sewing needles
- 1 disc magnet
- bowl of water
- cardstock piece 10 from kit
- 2 small slices of cork (or plastic milk bottle lids will also work)

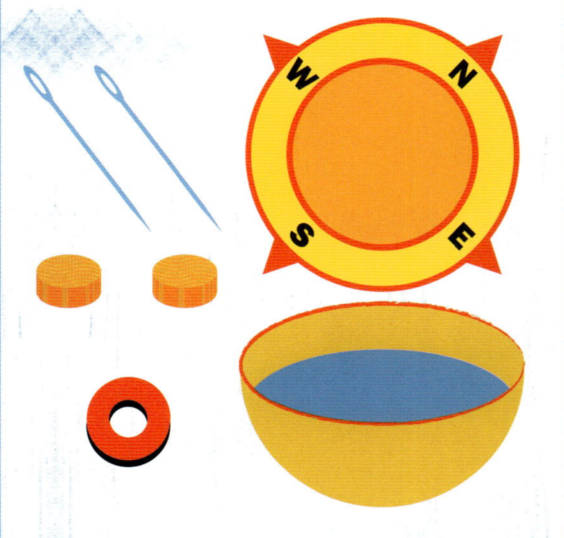

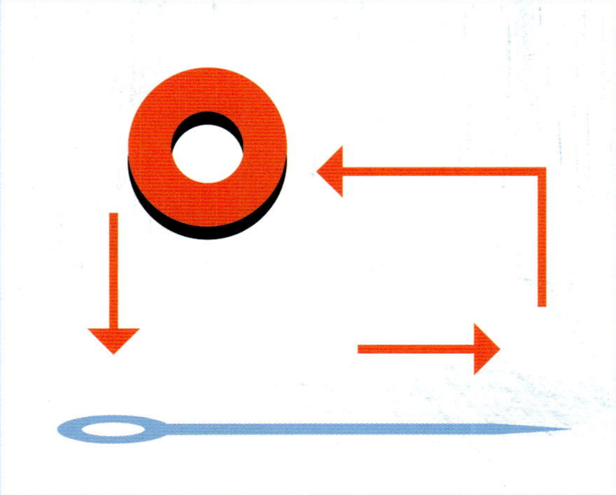

1 First, you need to magnetize the needles. Hold one needle by the eye and stroke the permanent magnet along it about thirty times. You always need to stroke from one end to the other, then start again at the first end. (If you just rub the magnet back and forth, it won't work.)

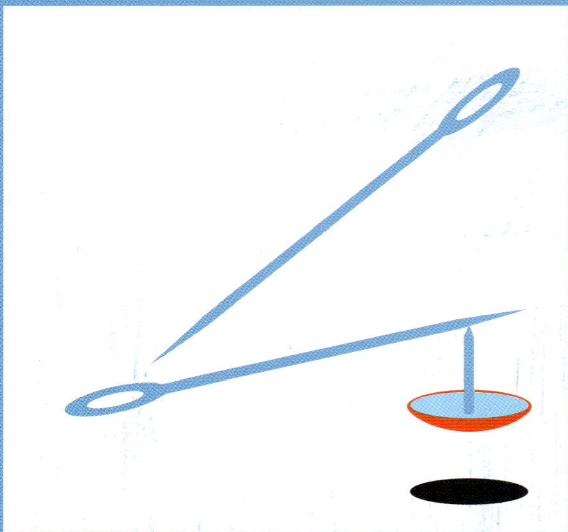

2 If you like, test your magnetized needles to see if they are strong enough to pick up a pin or another (non-magnetized) needle. If not, stroke them some more.

3 Now put piece 10 on a flat surface and put the bowl of water in the center of it.

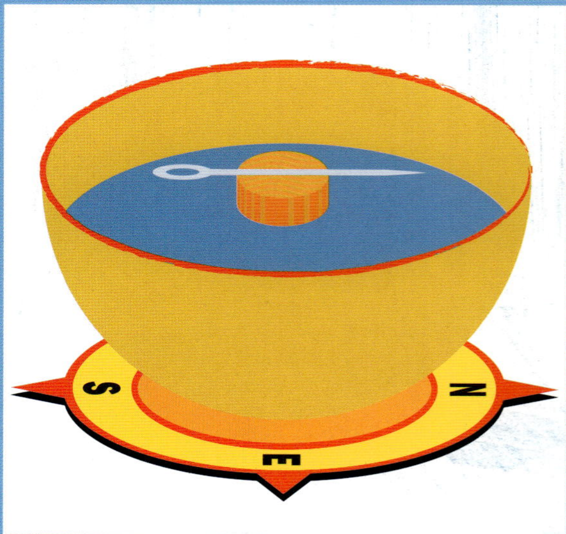

4 Float a piece of cork (or the plastic lid) in the water and carefully lay one of the magnetized needles on it. Wait for it to stop moving and then see which way the needle is pointing.

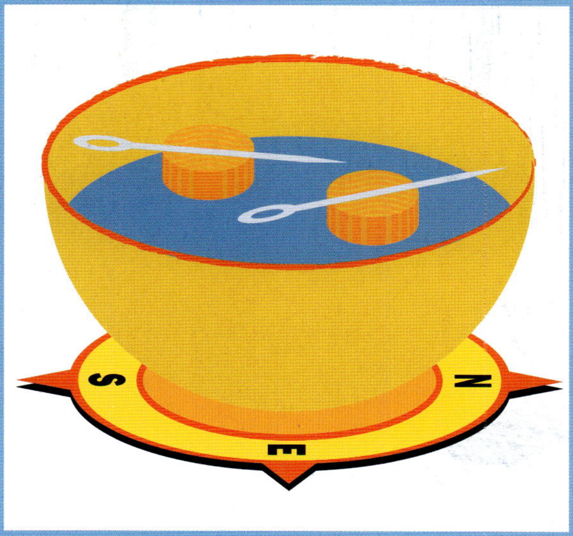

5 Repeat step 4 for the second needle.

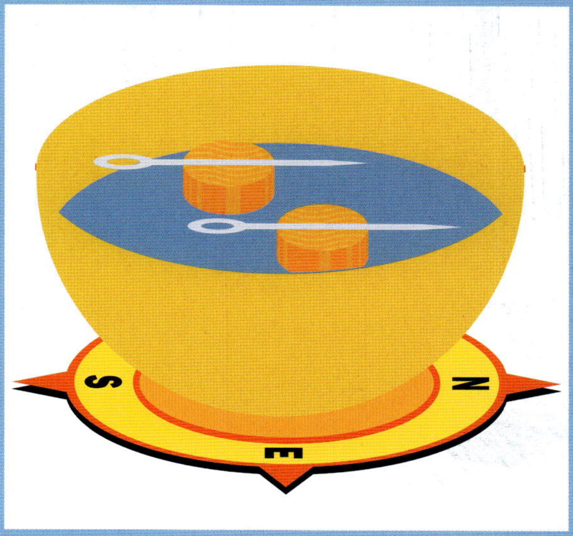

6 Wait for both the needles to point in the same direction. If the two needles don't point in the same direction, it may be because they are not floating freely, so check that they are. Another possible problem is that they may not be magnetized enough. Try stroking them with the magnet again—make sure you rub in the same direction as before!

7 One end of each needle is pointing north, and the other is pointing south. To find out which is which, go outside around midday when the sun is shining and look at your shadow. If you live in the northern hemisphere, your shadow will point north. If you live in the southern hemisphere, it will point south. Work out which end of the needles is pointing north.

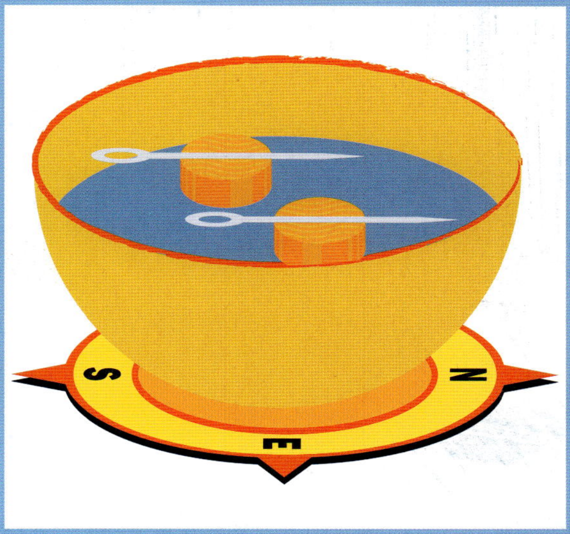

8 Turn card piece 10 under the bowl until the "N" lines up with the end of the needles that is pointing north, as you figured out using your shadow in step 7. You can now use your compass to find east, south, and west, too!

HINT

You made your needles into tiny magnets. If a magnet is allowed to move freely, it will line up with the Earth's magnetic field! Once it does, you can use it to find your way.

MAGNETISM IN SPACE

Earth's magnetic field does more than just help people find their way around. It is like a force field that protects our planet from harm. In fact, there would be no life on Earth without it!

THE SOLAR WIND

The Sun constantly sends out a stream of tiny charged particles. These particles flow out in all directions, traveling at very high speeds throughout the solar system. Part of this "solar wind" heads straight for Earth! It carries a serious dose of radiation that would be deadly for any living things on the surface. But luckily, the solar wind never reaches the surface. Earth's magnetic field deflects it like a shield. The force of the solar wind pushes the magnetic field out into a teardrop shape.

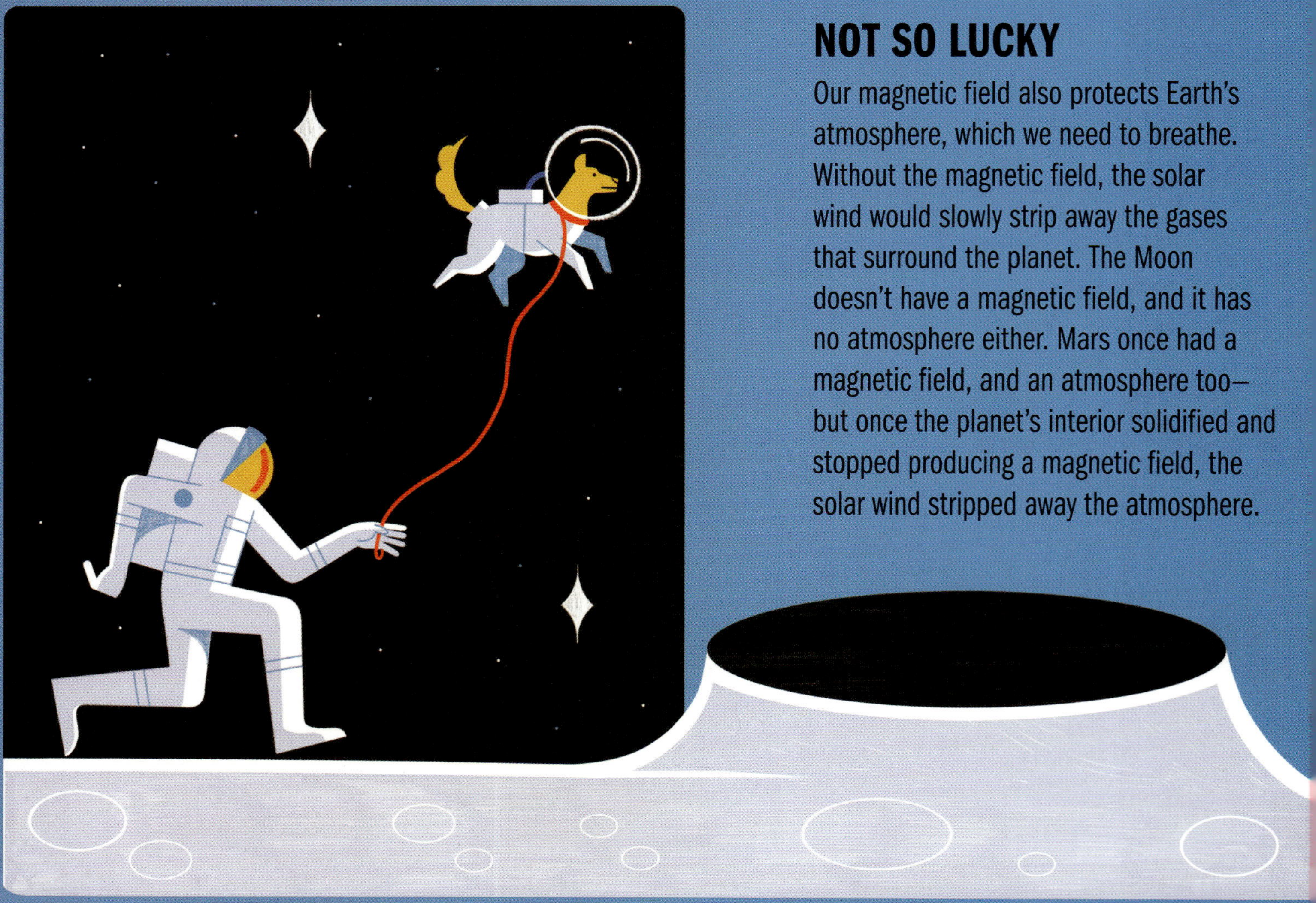

NOT SO LUCKY

Our magnetic field also protects Earth's atmosphere, which we need to breathe. Without the magnetic field, the solar wind would slowly strip away the gases that surround the planet. The Moon doesn't have a magnetic field, and it has no atmosphere either. Mars once had a magnetic field, and an atmosphere too—but once the planet's interior solidified and stopped producing a magnetic field, the solar wind stripped away the atmosphere.

STARS AND GALAXIES

Stars have magnetic fields. In fact, the Sun's magnetism—which is much stronger than Earth's—causes dark spots called sunspots and explosions of energy called solar flares. Astronomers have recently discovered that most stars the size of the Sun or larger have a magnetic field. Even galaxies have their own magnetic fields, though they are not always impressive. Our galaxy, the Milky Way, has a magnetic field that is much weaker than Earth's.

WEAK AND STRONG

Scientists use a unit called the gauss to measure the strength of a magnetic field. Earth's magnetic field is about 0.5 gauss, but it covers a huge area. A fridge magnet is often about 100 gauss, but its effects won't be felt over long distances. Scientists have made super-powerful magnets that have to be measured in units called teslas. A tesla is equal to 10,000 gauss.

HOME AND WORK

Big magnetic fields, like Earth's and those created by the magnets on a maglev train, can do big jobs. But even small magnets can be useful. You probably use magnets every single day, without even realizing it.

MAGNETS AT HOME

Most homes are full of magnets, and the fridge door is just a start. In addition to holding photos and notes to the door, magnets also hold the door closed. In fact, magnets are ideal for holding things closed. You'll find them on pet doors, notebook covers, jewelry clasps, and cupboard doors. There are also magnets in toys. The carriages on a toy train stick together using magnets, and travel checkers sets have magnets in the pieces to keep them from sliding around. You can even build with magnetic rod-and-ball sets!

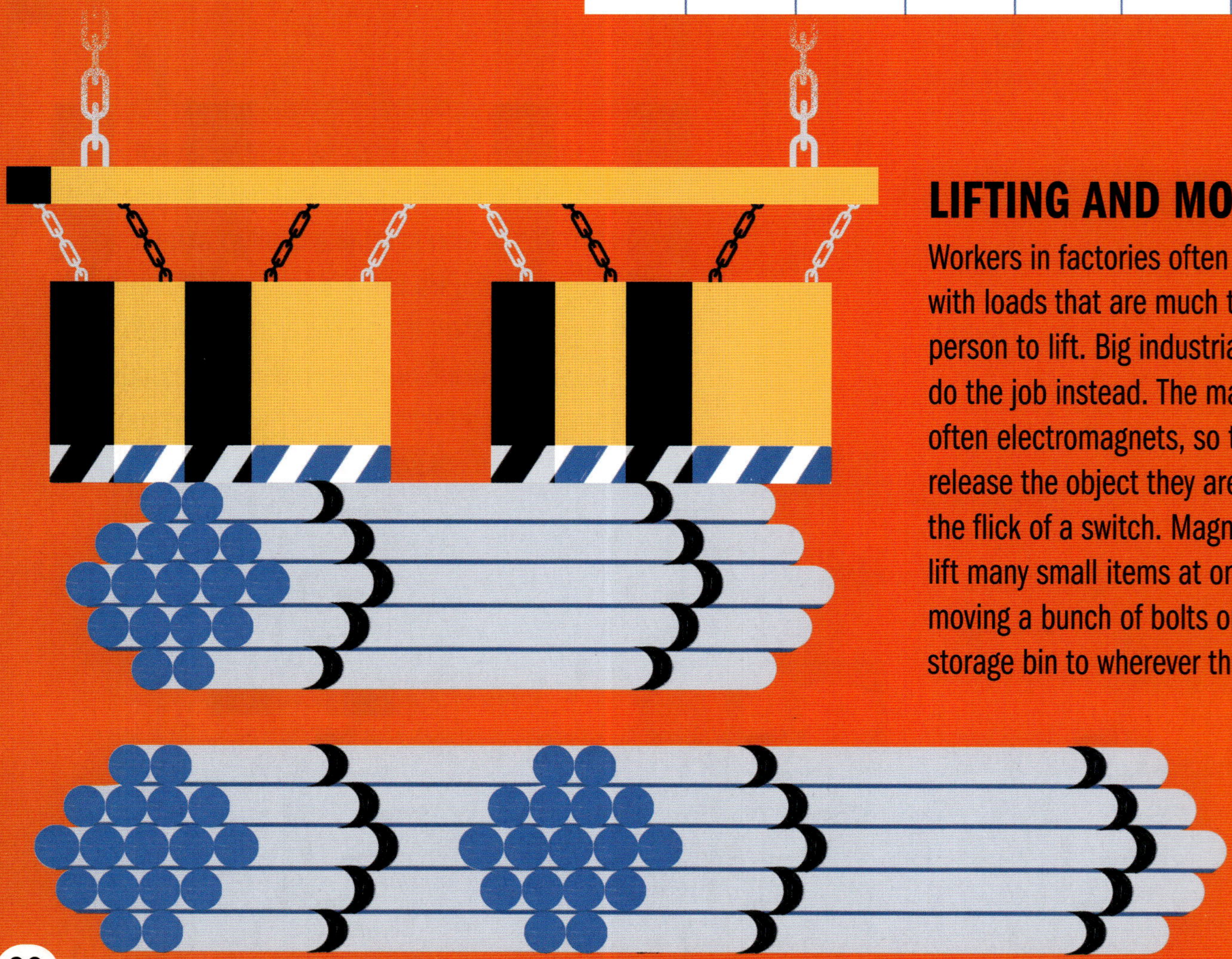

LIFTING AND MOVING

Workers in factories often have to deal with loads that are much too heavy for a person to lift. Big industrial magnets can do the job instead. The magnets used are often electromagnets, so that they can release the object they are carrying with the flick of a switch. Magnets can also lift many small items at once, such as moving a bunch of bolts or screws from a storage bin to wherever they are needed.

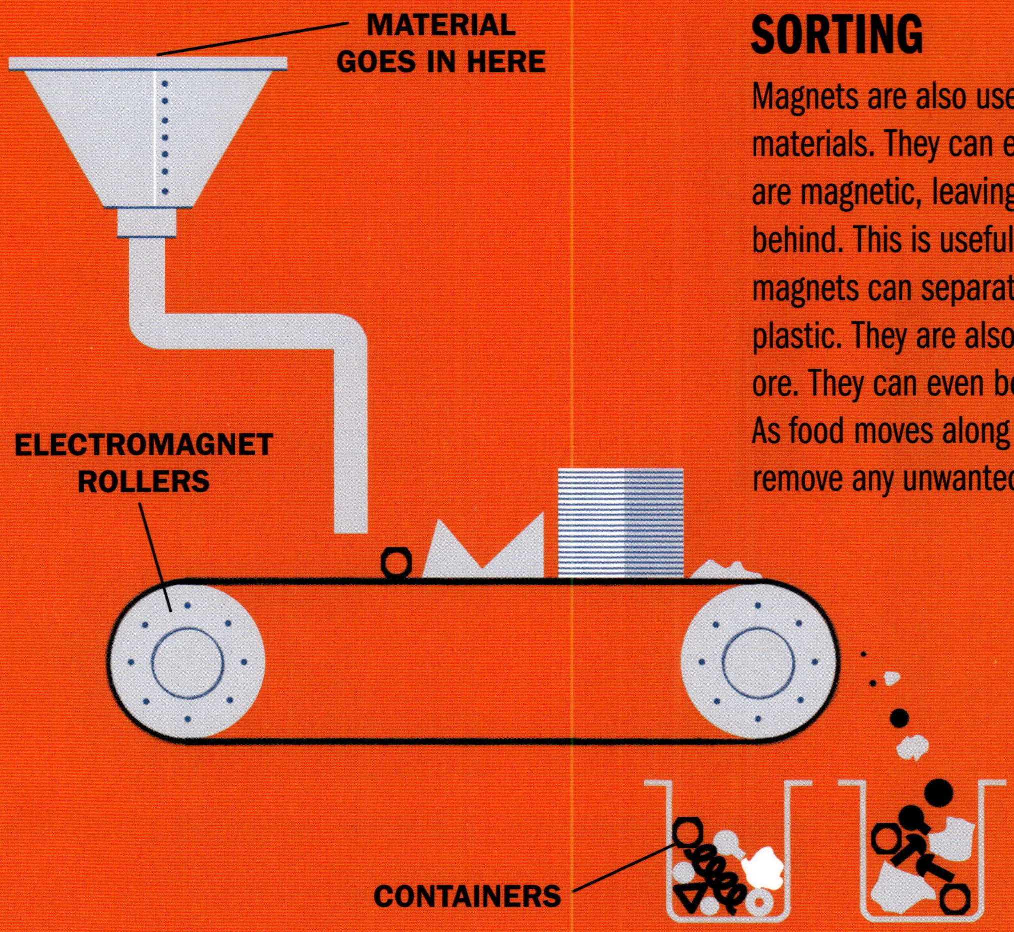

SORTING

Magnets are also useful for sorting and separating materials. They can easily pull out materials that are magnetic, leaving only non-magnetic materials behind. This is useful at recycling plants, where magnets can separate metal waste from paper or plastic. They are also used in mining to extract iron ore. They can even be used in food production! As food moves along a conveyor belt, magnets remove any unwanted metal contamination.

GET IT RIGHT!

A product designer has to choose the right strength of magnet for the job. For example, the magnet on a fridge door has to be strong enough to keep it closed. But it can't be too strong, or the fridge would be too hard to open.

MAKING MAGNETS

Magnets are not just used in factories—some of them are made there! Different metals are heated until they melt and mix. Once they cool and become solid, they are crushed into powder. The powder is put into a mold and a magnetic force is applied by powerful machines that press the powder into shape. Then the new magnet is heated up so that the powder particles fuse together.

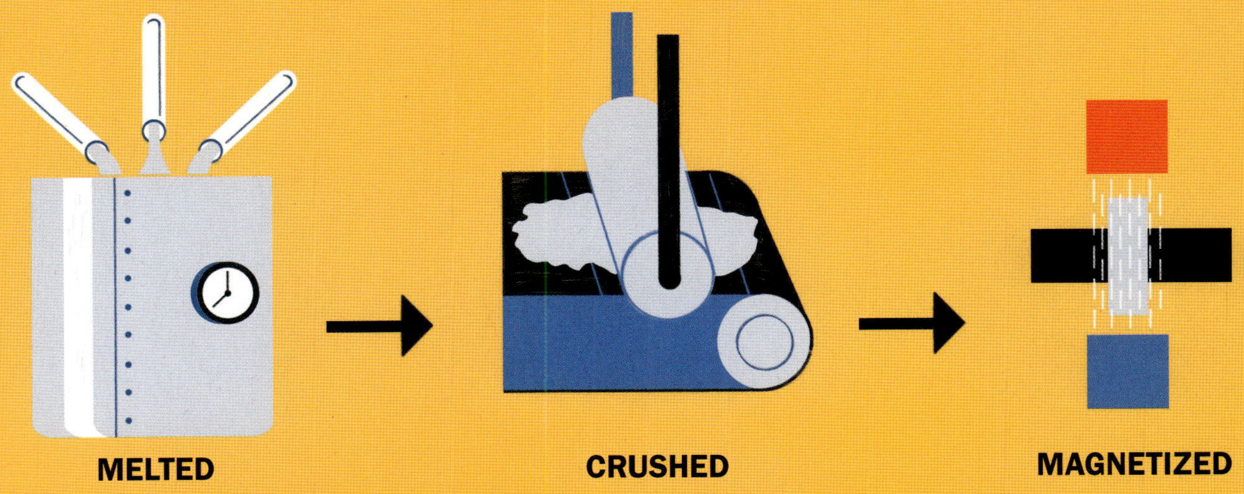

BUILD A CRANE

You've seen how magnets can be used at work. Now build your very own magnetic working machine.

YOU WILL NEED:

- 1 paperclip
- 1 disc magnet
- cardstock pieces 3–8, 11, and 12 from kit
- string
- tape

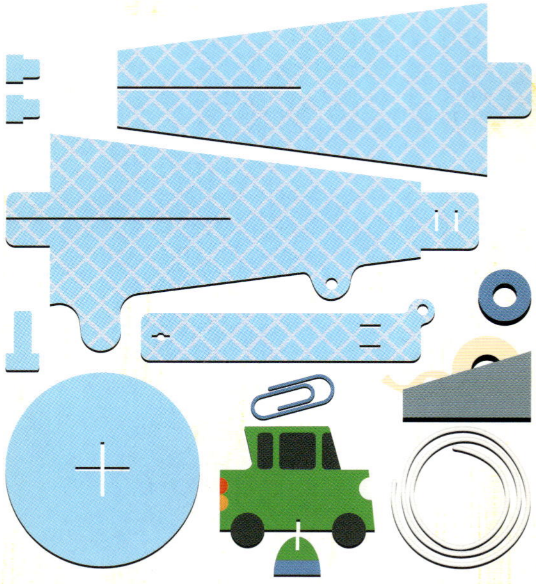

1 Slide piece 3 into piece 4.

2 Line up the two slots on piece 5 (the crane arm) with the two slots on the top of piece 3. Use pieces 6 and 7 to secure piece 5 to piece 3.

3 Take the crane half of the tray and push out the circle with the cross. Push the bottom of pieces 3 and 4 into the hole in the tray. Secure underneath with piece 8.

4 Stand the crane upright and check that you can swivel it left and right while the tray stays still.

5 Loop the string through the disc magnet and tie in place.

6 Take piece 11 from the kit and tape the paperclip in place, as shown.

7 Push piece 11 into the slot on piece 12, to make your car stand up on its own.

8 Push the other end of the string through the hole on the end of the crane arm (piece 5), then through the hole on the back of the crane arm. Finally, push the string through the hole on piece 3.

9 Use the tab on the back of piece 3 to rotate your crane. Lower the disc magnet, using the string, and try to pick up the car.

HINT

The magnet on the crane is attracted to the paperclip on the car. The force of their attraction is strong enough to overcome gravity and lift the car into the air. A real crane like this would use a type of magnet called an electromagnet, which can be turned on and off. This would allow the crane to drop the car where it needs to go.

MAGLEV EXPRESS

When a train rolls along the tracks, friction on its wheels slows the train down. But what if you could invent a train that didn't touch the tracks at all? A maglev, short for "magnetic levitation," uses the force of magnetism to lift itself off the rails. With no friction, it can zoom along at enormous speeds. The record is 373 miles (600 km) per hour!

WHERE ARE THEY?

Magnetic train technology was first developed in Britain, Germany, and Japan, but the first high-speed line was built in China. It connects Shanghai airport to Shanghai city center. The train has a top speed of 268 miles (431 km) per hour.

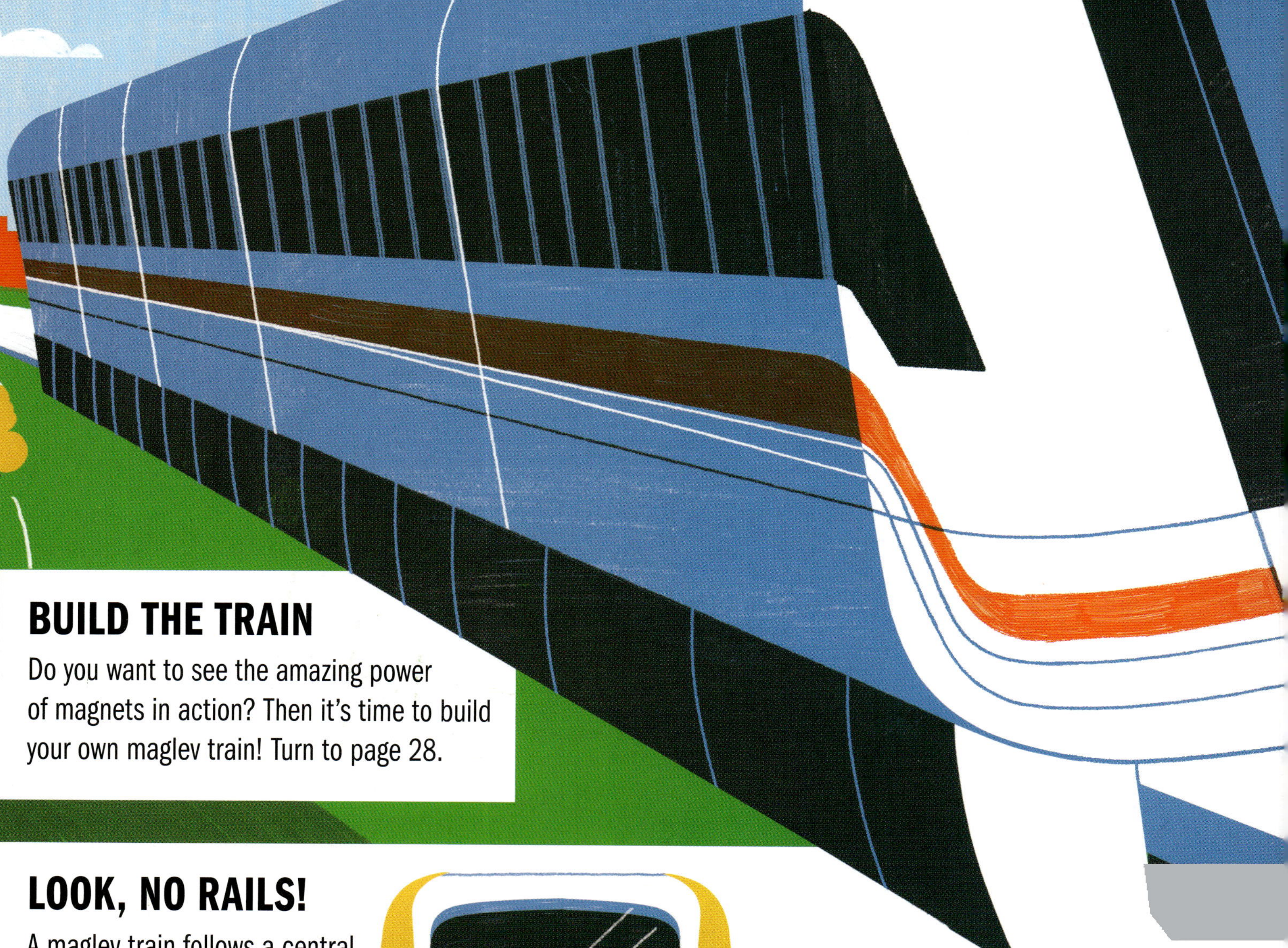

BUILD THE TRAIN

Do you want to see the amazing power of magnets in action? Then it's time to build your own maglev train! Turn to page 28.

LOOK, NO RAILS!

A maglev train follows a central beam called a guideway. Magnetic field controls lift the train ½ inch to 4 inches above the beam. They also pull the train forward and can make it slow down or stop.

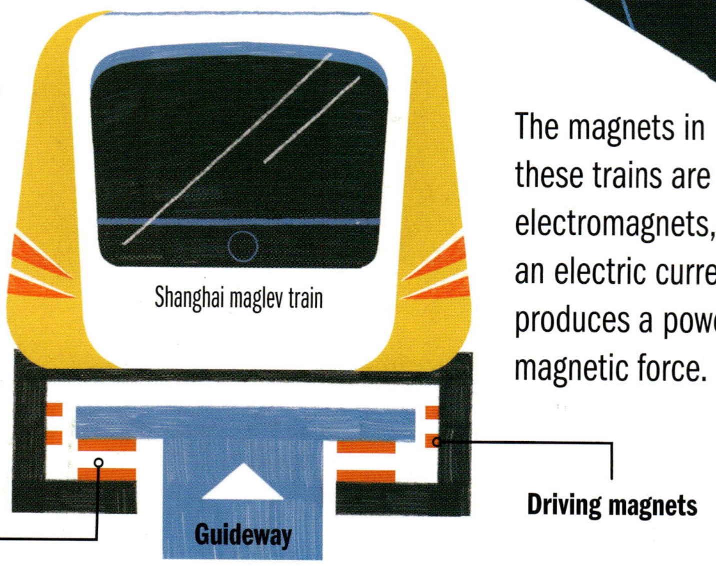

The magnets in these trains are electromagnets, where an electric current produces a powerful magnetic force.

CLEAN AND GREEN

Maglev trains are quieter than diesel or electric trains. They have no engines to produce waste gases that pollute the air.

VACUUM TUBE TRAINS

Maglev trains have no friction from the tracks, but they do face resistance from air and wind as they whiz past, and this slows them down. Engineers think that if maglev trains drove through tubes from which the air had been reduced or removed, they could reach incredible speeds, perhaps 5,000 miles (8,000 km) per hour!

MAGLEV TIMELINE

- **1967** Gordon Danby and James Powell receive a patent for a maglev system.
- **1979** A working maglev train is exhibited in Hamburg, Germany.
- **1984** The world's first commercial maglev train opens in Birmingham, UK, taking passengers from the airport terminal to the train station.
- **2004** The Shanghai maglev, the world's fastest passenger train, begins operations.
- **2015** A new maglev service opens at Incheon Airport in South Korea.

MAGNETS IN TRANSPORT

You already know that magnetism can power a train. After all, you're about to build your own working model! But there are also magnets in many other forms of transportation—even spacecraft.

GOODBYE, GAS

People around the world are trying to use less gasoline and diesel fuel for transport. When vehicles burn these fuels, they spew gases such as carbon dioxide into the air. Large amounts of carbon dioxide build up in the atmosphere where they trap the sun's heat, like the glass in a greenhouse. This causes climate change. Getting gasoline and diesel vehicles off the road is one way of fighting climate change. Many of them are already being replaced by electric cars. And what's inside an electric motor? Magnets, of course!

MAGNETS IN CARS

Although they don't have an electric motor, gasoline and diesel vehicles are still full of magnets. There are magnets in an anti-lock braking system, magnets to tell if a passenger's seat belt is clipped in, and magnets that alert the driver if a door is open. The meters that measure a car's speed and engine revs use magnets, too.

TRAINS AND SHIPS

Electric trains have magnets in their motors, in a similar way to electric cars. There are also magnets in many ships, and sailors can even use magnetic patches to cover holes in a vessel's metal hull. The magnet creates a water-tight seal until the damage can be repaired. There are magnets in aircraft too, from the loudspeakers that carry the captain's messages, to fuel pumps, speed sensors, and flight controls.

MAGNETS IN SPACE

Astronauts on the International Space Station have done experiments with magnets to prove that they work just as well in the weightless conditions of space as they do on Earth. In fact, magnets were used during the Apollo 11 Moon landing in 1969. The spacecraft's memory system generated power using magnets instead of electricity. Today, magnets inside satellites help them keep a stable position in orbit. They are also used in robotic spacecraft, and one day they may be part of a magnetic system for powering space flight.

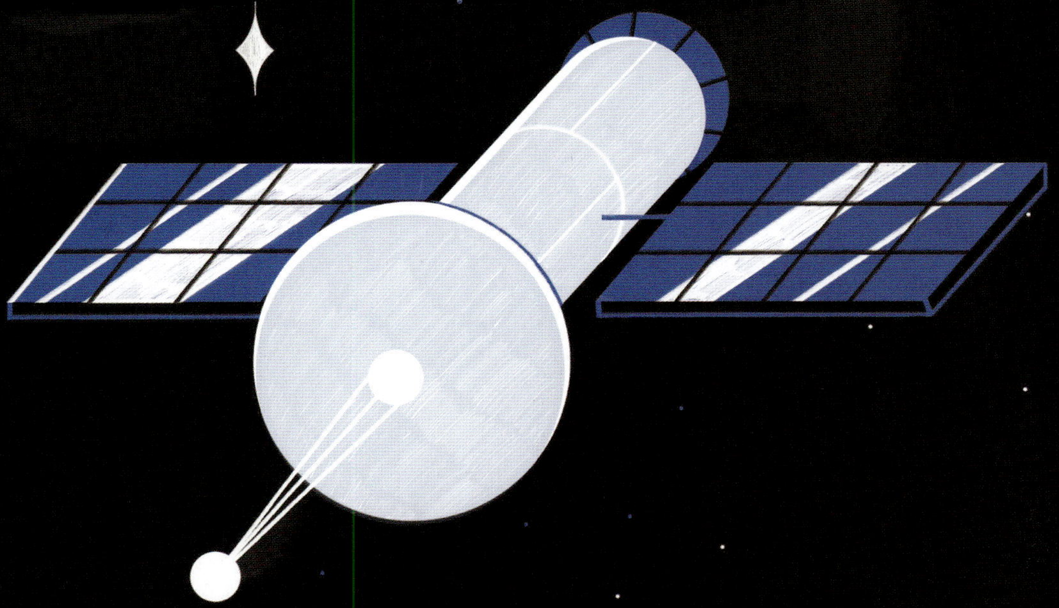

BUILD A MAGLEV TRAIN

Using the force of magnetism, it's now time to build your very own working maglev train! Follow these instructions and get started.

YOU WILL NEED:

- 2 magnetic strips
- 2 disc magnets
- tape
- cardstock pieces 13–27, 47, and 48

1 Attach pieces 13a and 13b together. Cut the magnetic strips into lengths of 11.4 inches. Peel off the backing from two of the magnetic strips, and stick them to the length of pieces 13a and 13b.

2 Slot the tabs on pieces 13a and 13b into the slots on pieces 14a, 14b, 15a, and 15b. Next, slot pieces 16 and 17 into the slots on each end of 14a/b and 15a/b. These are the buffers.

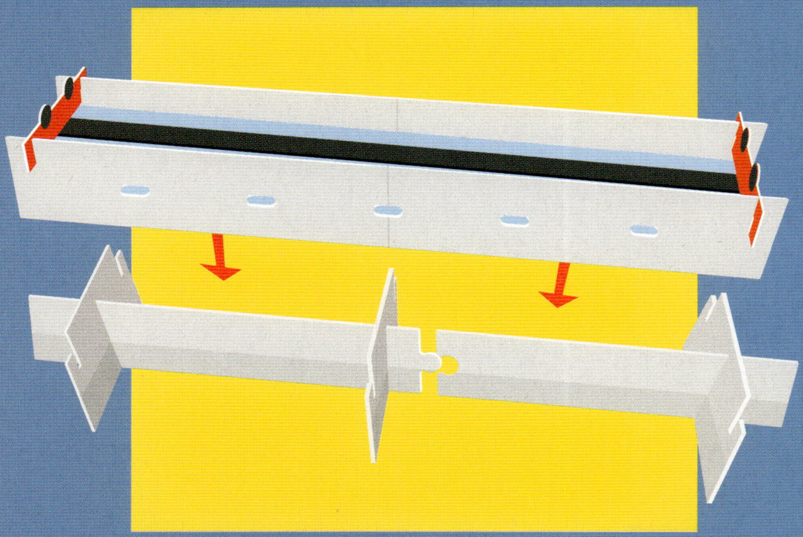

3 Slot pieces 19, 20, and 21 into piece 18a and b, as shown. Then take the track section (13-17) and slot it on top of pieces 19, 20, and 21.

4 Stick pieces 22 and 23 together with tape. Peel off the backing from the last two lengths of the magnetic strip. Stick the strips to the outer edges of 22/23, as shown.

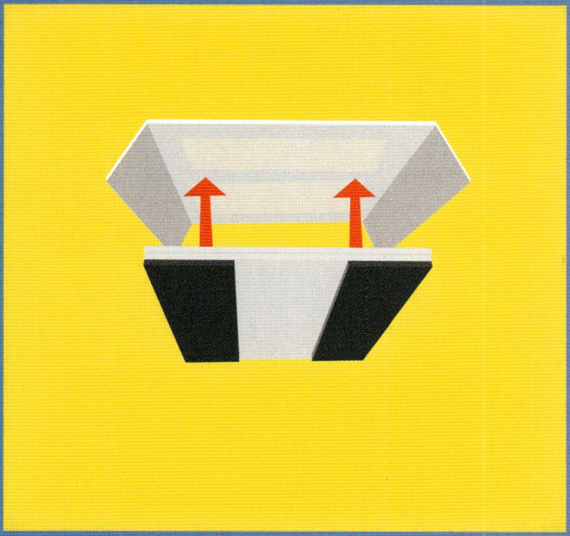

5 Take piece 24 and fold down the sides. Stick piece 22/23 in the gap between the folded sides. Make sure the magnets are facing down.

6 Take piece 25 and stick the folded part of piece 24 to its bottom edge, with tape.

7 Stick piece 26 to the other side of piece 24 in the same way. Then slot piece 27 into pieces 25 and 26, as shown.

8 Your completed maglev train engine should look like this.

9 Place your train onto the track. It should float in place and if you push it, it should move along.

10 Now place a disc magnet in the back of the train and use a piece of tape to hold it in place. It does not matter which way up you tape it on.

11 With the train on the track, take the other disc magnet and build the magnet wand from page 43, step 5, using pieces 47 and 48. Position the wand behind the disc magnet on the train. You should be able to power your train down the track by using the magnets. Just remember you need the magnets to repel, so make sure you have your magnet the right way up.

12 You can use the top of the box that the pieces came in, to stand your maglev on.

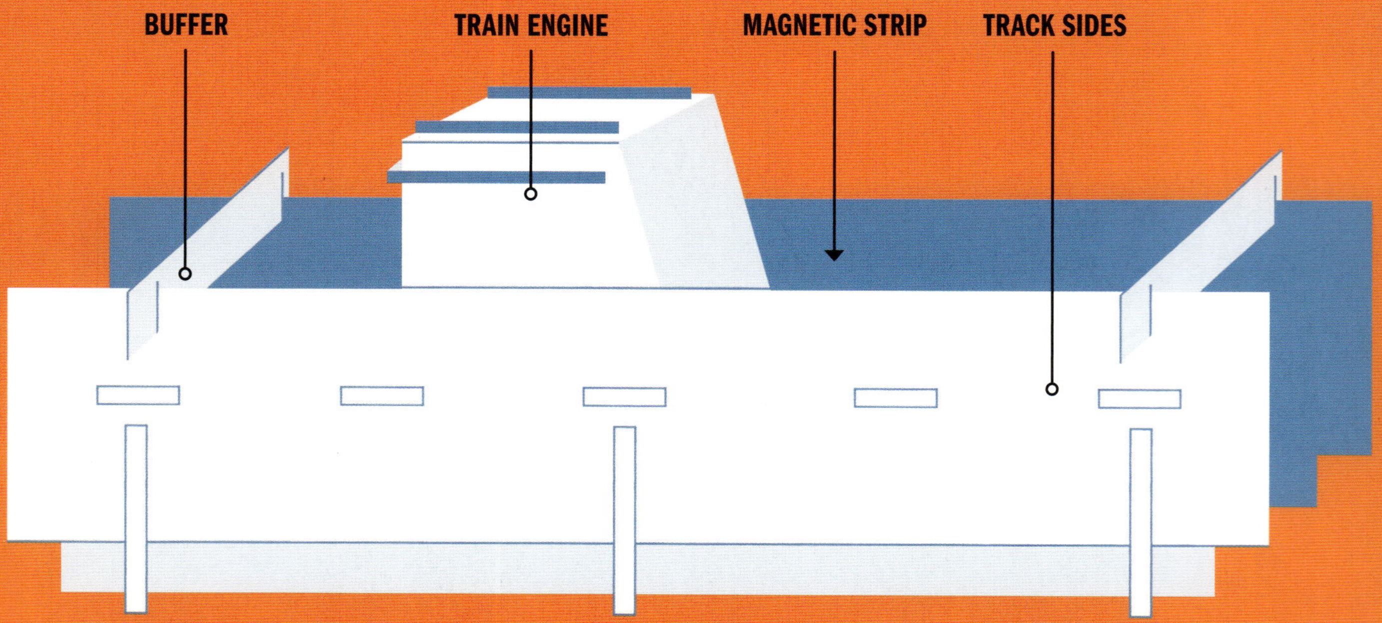

HOW YOUR MAGLEV WORKS

Maglev trains need really powerful magnets. That's why they use electromagnets that are kept at very low temperatures to make them even stronger. Your train uses permanent magnets, but the basic idea is still the same.

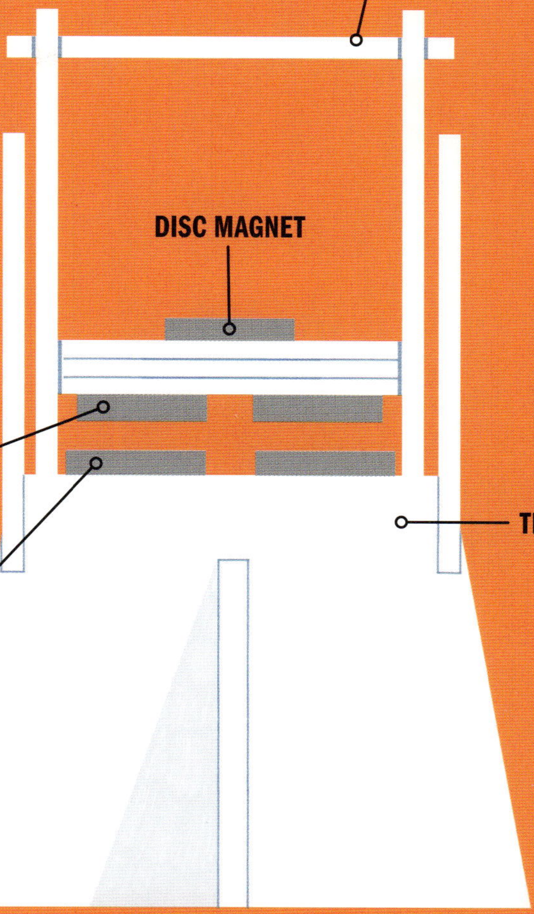

PROPULSION

In a real maglev train, electromagnetic coils in the side of the track create magnetic fields that push and pull the train along. Your train doesn't have these. Instead, you use the force of the disc magnets repelling each other to push it along.

The driving magnets pull the train forward, and can make it slow down or stop.

LEVITATION

Your maglev train works because the strip magnets on the underside of the train repel the strip magnets on the tracks. Poles of the same type (for example, two north poles) repel each other, so if the same kind of poles are facing each other, the train will levitate just above the track.

ELECTRIC MOTORS

There are electric motors in toys, toothbrushes, washing machines, and cars. They turn electrical energy into movement energy, and they rely on the power of magnetism to make it all work.

FROM ELECTRICITY TO MOVEMENT

The first electric motor was invented by Michael Faraday in the 1820s. A Danish scientist had recently discovered that a compass needle moved when the electric current from a battery was switched on and off. Faraday realized that this phenomenon could make it possible to use electricity to create motion. He built a device that rotated when supplied with an electrical current.

HOW IT WORKS

An electric motor makes use of a magnet's ability to repel another magnet. By forcing two magnets to push against each other, it creates a spinning motion that can be used to do anything from rotating a toothbrush head to blending up a smoothie. Most motors use electromagnets. Electromagnets are made of coils of wire with electricity passing through them. Moving electricity creates magnetic fields, so the coils behave like a magnet when electricity is flowing.

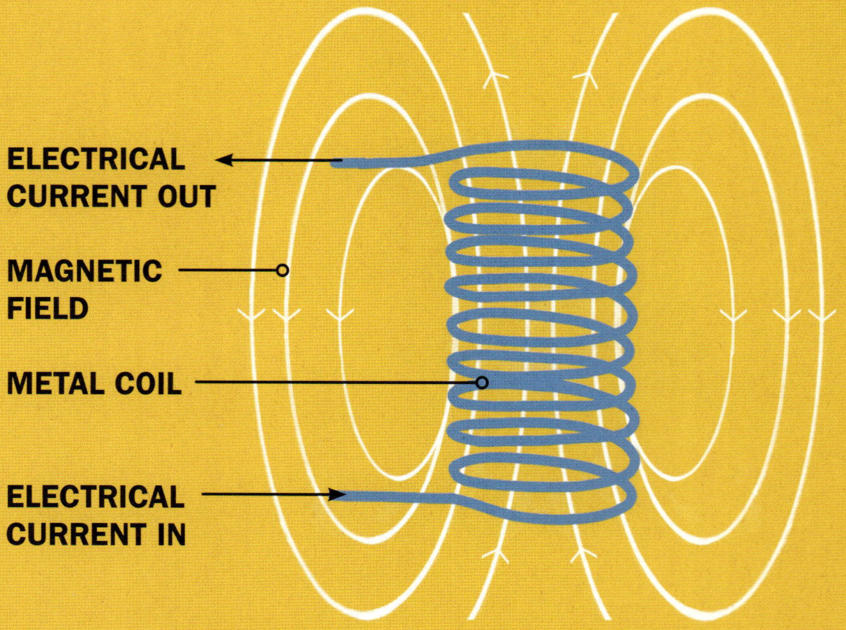

WASHING MACHINE

FOOD MIXER

JUICER

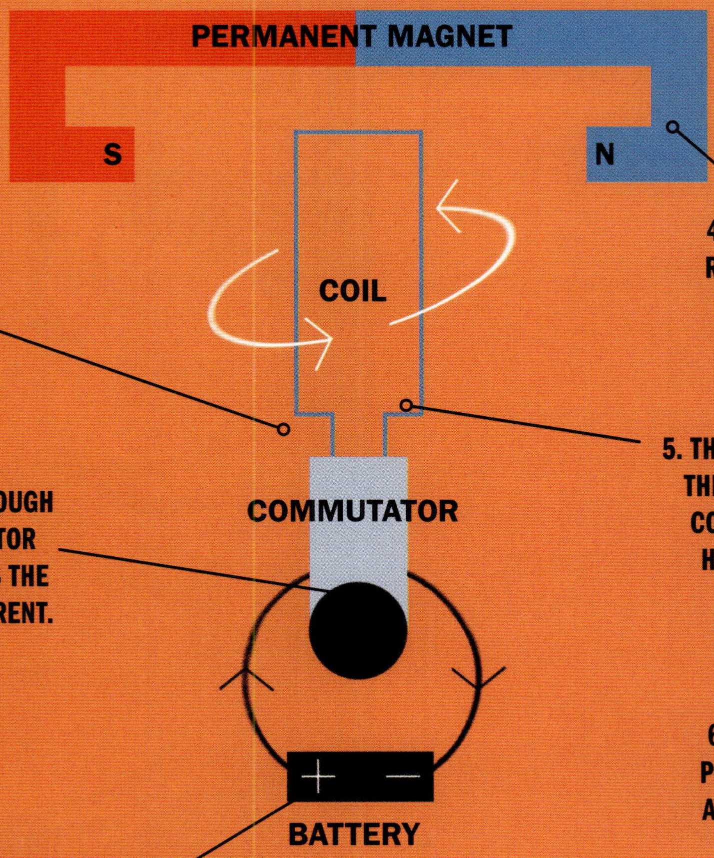

1. IN ORDER TO WORK, AN ELECTRIC MOTOR MUST BE CONNECTED TO A POWER SOURCE SUCH AS A BATTERY.

2. THE ELECTRICITY TRAVELS THROUGH THE COMMUTATOR. A COMMUTATOR IS LIKE A SWITCH THAT REVERSES THE DIRECTION OF THE ELECTRIC CURRENT.

3. THE ELECTRICITY THEN TRAVELS INTO THE COIL, TURNING IT INTO AN ELECTROMAGNET WITH A NORTH AND SOUTH POLE.

4. THE NORTH POLE OF THE COIL REPELS THE NORTH POLE OF THE PERMANENT MAGNET, WHICH MAKES THE COIL TURN.

5. THE COMMUTATOR THEN REVERSES THE CURRENT, WHICH MAKES THE COIL TURN THE OTHER WAY. THIS HAPPENS AGAIN AND AGAIN AT SPEED.

6. THE MOVEMENT IS USED TO POWER THE MACHINE, SUCH AS AN ELECTRIC TOOHBRUSH OR A BLENDER.

REMOTE-CONTROLLED CAR

FAN

VACUUM CLEANER

MAGNET MAGIC

BOUNCING MAGNETS

Magnetism can sometimes seem a bit like magic. And you can use this force to create some seriously spooky effects!

YOU WILL NEED:

- 2 disc magnets
- piece of printer-size paper
- cardstock pieces 28 and 29 from the kit
- tape

1 Slot piece 28 into piece 29, as shown.

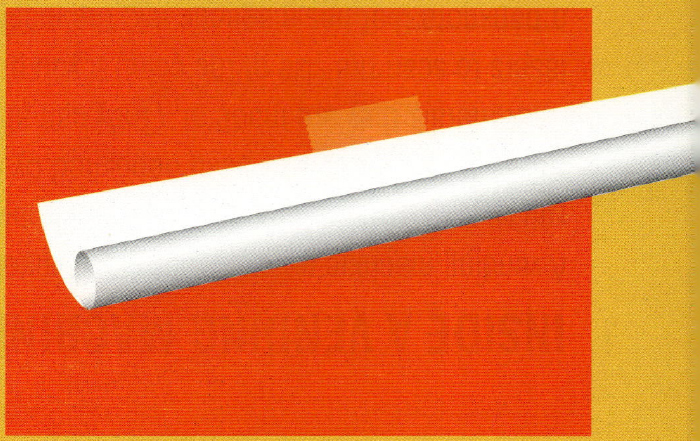

2 Roll the piece of paper into a tube about the width of a pencil. Secure the tube with tape. Make sure your tube of paper fits easily into the hole of your disc magnets.

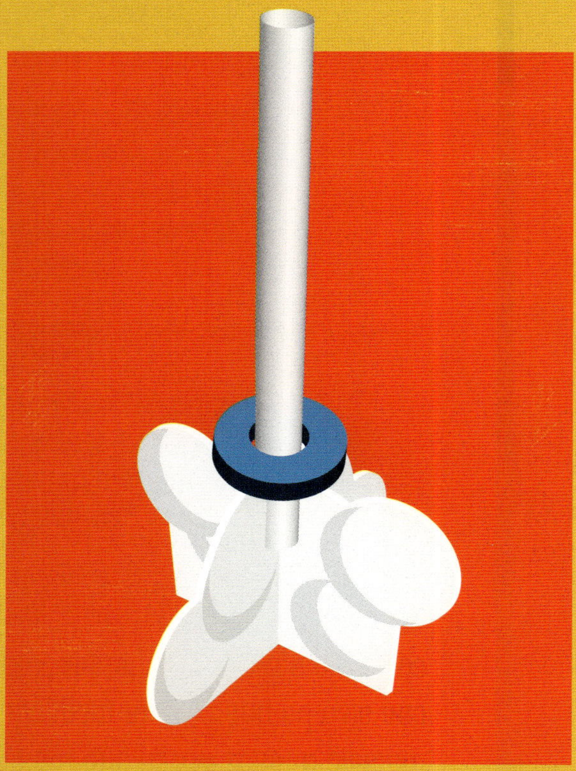

3 Push the paper tube into the slots on pieces 28 and 29, so it is held upright.

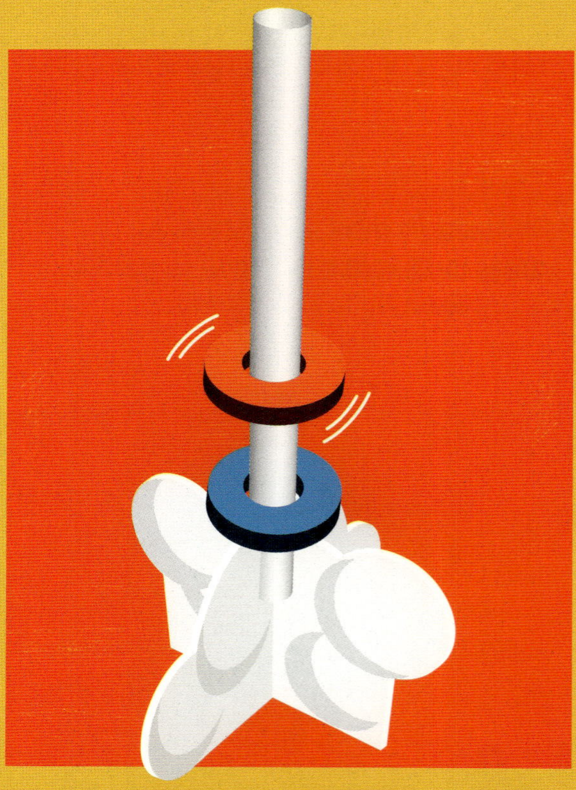

4 Drop one disc magnet down onto the paper tube, until it reaches the bottom. Push the other disc magnet onto the tube in the same way. Make sure you have the same poles facing each other. It should float just above the first disc magnet. What happens when you push it down slightly and then let go? What about if you slide the lower magnet up the tube?

HINT

The same poles of the two disc magnets repel each other. Just like magnetic attraction, repulsion works at a distance and can overcome gravity. The upper magnet and lower magnet are repelling each other, so it bounces and resists if you try to push them together.

MAGNETIC PENDULUM

A pendulum is a weight that hangs from a line and swings back and forth. Normally, pendulums move in a very predictable way, but when you add magnets to the mix, it all goes a bit strange...

YOU WILL NEED:

- 2 disc magnets
- string
- cardstock pieces 3–8 from the kit

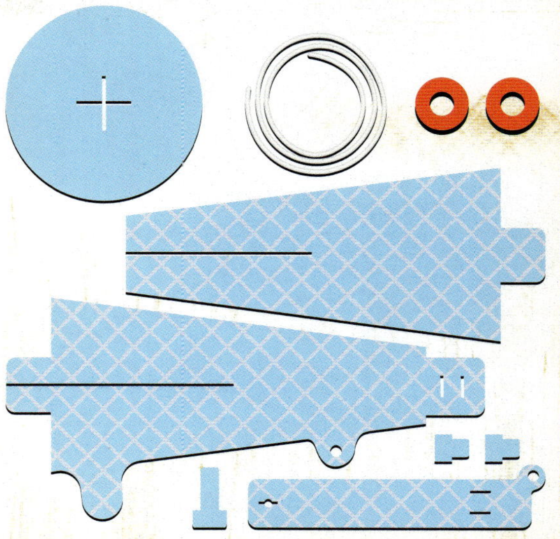

1 Build the crane arm, using the instructions on page 22.

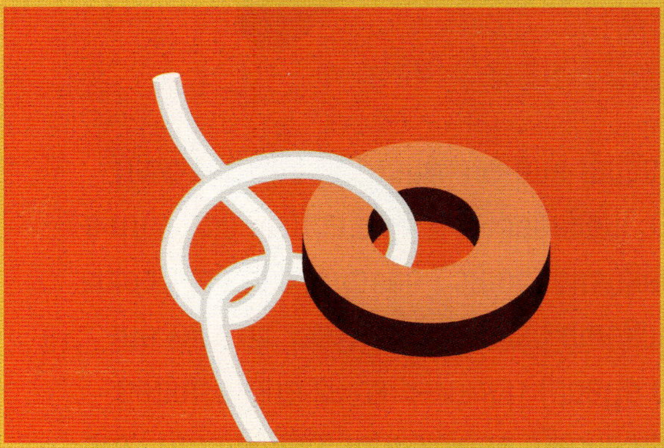

2 Tie one end of the string to one of the disc magnets.

3 Loop the other end of the string through the holes on the crane (see page 23).

4 Place the other disc magnet on the tray. It should be directly below the magnet that is hanging down.

5 Swing the string and magnet in any direction and watch what happens.

HINT

The magnet on the string behaves very strangely. One side of it is attracted to whichever pole of the magnet on the base is facing upward. The other side of the hanging magnet is repelled by it. As the string swings, both of these forces are at work, making the magnet move about.

MAGNETS AND SOUND

Magnets can run motors and generate electricity, but they do even more than that! Magnets are a key part of the speakers found in gadgets from stereos to smartphones.

SOUND WAVES

Sound is produced when an object vibrates. The skin of a drum vibrates when you hit it, and your vocal cords vibrate as you push air past them. These vibrations produce invisible waves that travel through the air. When you speak, the sound waves that you produce travel through the air and into the ears of your listener.

SENDING SIGNALS

But what if you want to speak to someone in a different city over the phone? Sending sound waves through the air won't work. You need to convert your voice into an electric signal that can be sent over the phone network. When your friend's device receives the signal, it uses a loudspeaker to convert that electrical signal back into sound waves.

PARTS OF A SPEAKER

Loudspeakers of different sizes are found in stereo speakers, phone handsets, and even earbud headphones. No matter what the casing looks like, they all have the same three basic parts: a cone, a permanent magnet, and an electromagnet.

INSIDE A SPEAKER

Just like in an electric motor, an electrical signal changes the direction of the electromagnet coil's magnetic field, which makes the coil vibrate within the speaker. It swaps back and forth between being attracted to the permanent magnet and repelled by it. The cone is made of a flexible material attached to the electromagnet. It amplifies the vibrations of the electromagnet to produce sound we can hear.

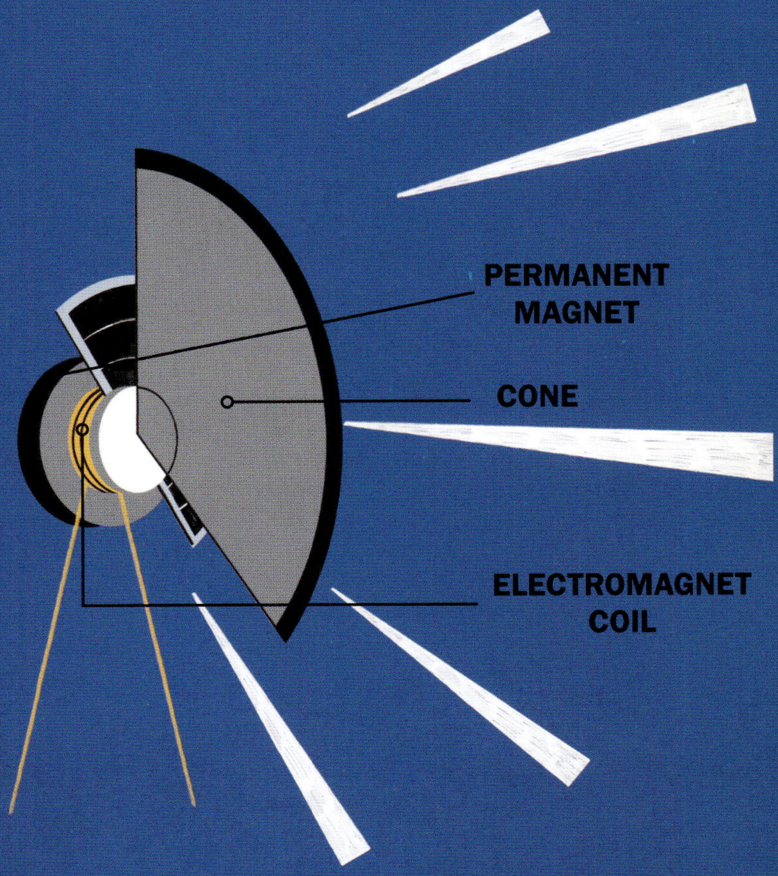

STOP THIEF!

There are also magnets inside the anti-theft tags that set off an alarm when you try to leave a shop without paying for something. Each tag has a thin strip of permanent magnet and another thin metal strip, which vibrates when placed in a magnetic field. The archway of the shop door has a transmitter in one side and a receiver in the other. Sending a signal makes the metal strip vibrate and disrupt the magnetic field. The receiver detects this and triggers an alarm.

MAGNETS AND MONEY

You might not realize it, but magnets and money go hand in hand. In fact, whenever you buy something from a vending machine, you are using the power of magnetism!

MAGNETIC OR NOT?

Every country makes its own coins, and they use different mixes of metals. Some of these are magnetic, and some are not. For instance, coins in the United States are made from copper, nickel, and zinc. They won't stick to a magnet. (Nickel is magnetic, but US coins have too small a percentage of nickel for a magnet to have an effect.) However, coins in the United Kingdom have a lot of steel, so they are magnetic. Some Canadian and Euro coins are magnetic, too. Test some coins with a magnet and see what you can discover!

SPOTTING FAKE COINS

Many vending machines use magnetism to identify counterfeit coins. They produce something called an eddy current, which is a loop of electrical current that can slow down magnetic objects as they fall. Because they are falling more slowly, the magnetic objects can be diverted into a separate chute from the non-magnetic objects.

CLOSE UP OF MAGNETIC STRIP
INSIDE A BANK NOTE

BANKNOTES

Criminals would love to be able to print their own counterfeit money. Governments add special features to banknotes that are hard to copy, to help people tell genuine notes from fake ones. These features include magnetic materials! Some banknotes have metallic threads woven into the note. They can also be printed with magnetic ink, which comes in a range of different colors.

INSIDE A VENDING MACHINE

A vending machine needs to know how much money you've put in. Some of them use electromagnets to tell what type of coins have been deposited by checking what metals they are made of. (They can also use lasers to scan the size of the coin.) If you pay with a banknote that has magnetic ink, the machine can "read" how much iron is in the ink. This proves whether a note is genuine, and what denomination it is. Once the money is counted and accepted, it's time to dispense your snack!

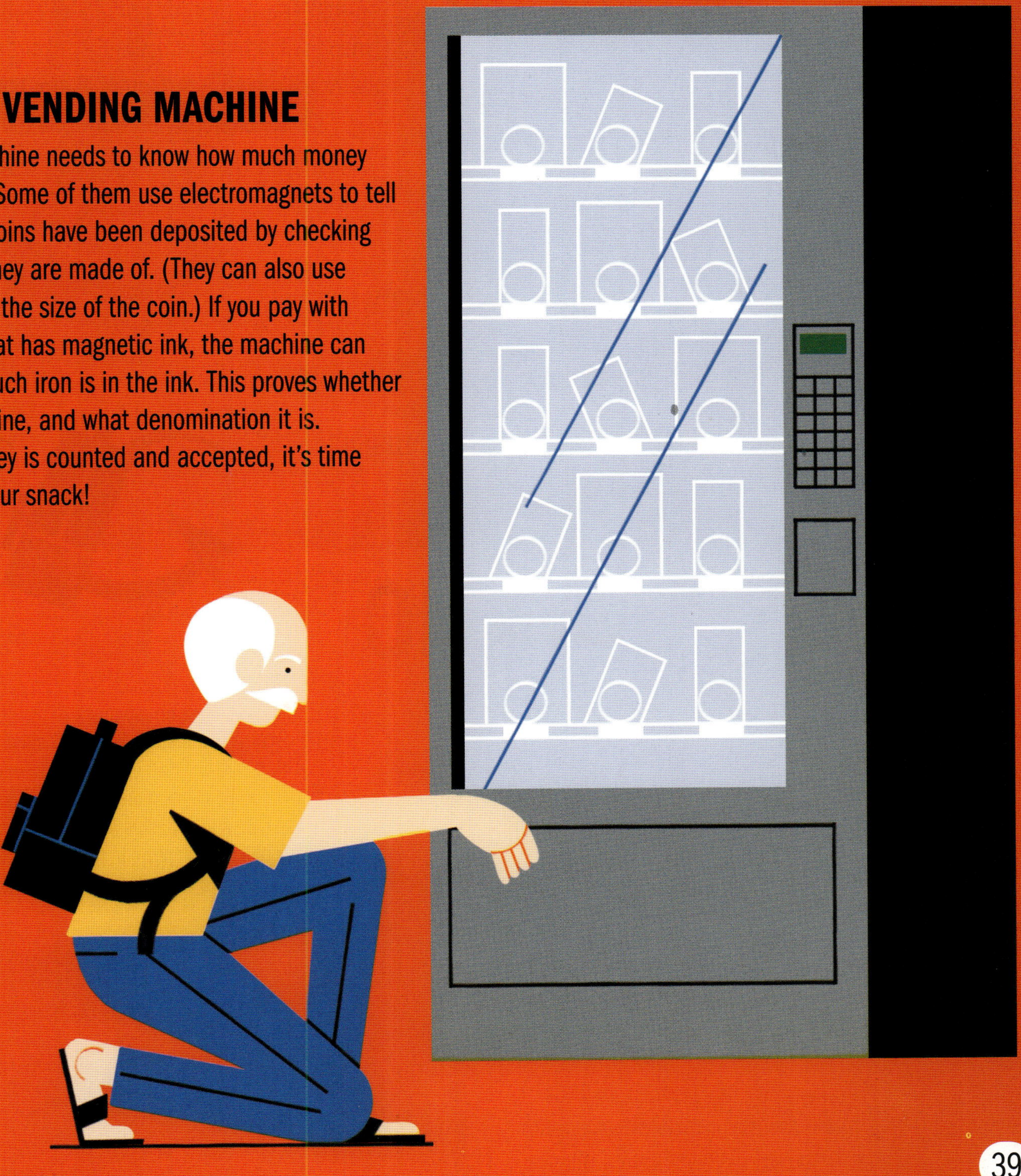

MEDICAL MAGNETS

Can magnets help diagnose and cure disease? It may seem like a crazy idea out of a science-fiction book, but the truth is that doctors and medical scientists use magnets every day.

SCANNING THE BODY

The most common use of magnets in medicine is inside an MRI machine. MRI stands for "magnetic resonance imaging," and it is a way for doctors to peer inside the human body without cutting it open. An MRI machine can produce detailed images of the body's organs, tissues, and bones. Unlike an X-ray or a CT scan, using an MRI does not expose the patient to radiation, which can be harmful in large doses.

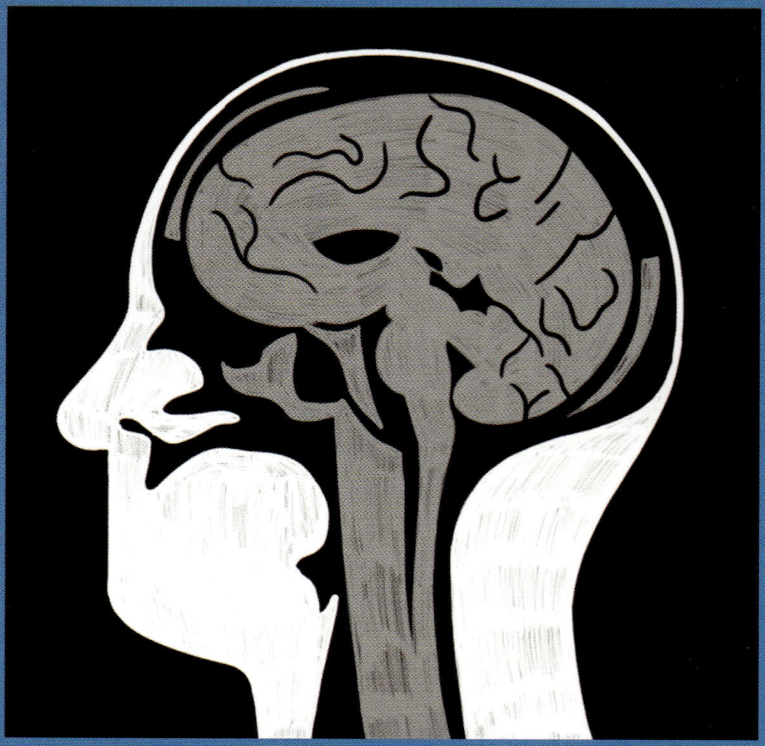

MRI IMAGE

HOW IT WORKS

For an MRI scan, a patient is placed inside a large, very powerful electromagnet. It produces a magnetic field of up to 2 teslas, or 20,000 gauss. Normally, producing a field this strong would take a huge amount of energy. To make it more efficient, the wires in an MRI's electromagnet are cooled to -452.4°F (-269.1°C). When a patient is placed in this magnetic field, it forces tiny particles called protons in their body to align with the field. Pulsing a frequency through the patient's body triggers sensors that can "read" the changes and turn them into an image.

TOO BIG?

The first MRI scan of a human body was made in 1977. The machine's inventor, Raymond Damadian, volunteered to be the test subject. But when he squeezed into the narrow tube and switched it on, there was no signal. He was just too big for it to work! One of his assistants, who was much thinner, got in the machine instead, and this time it worked.

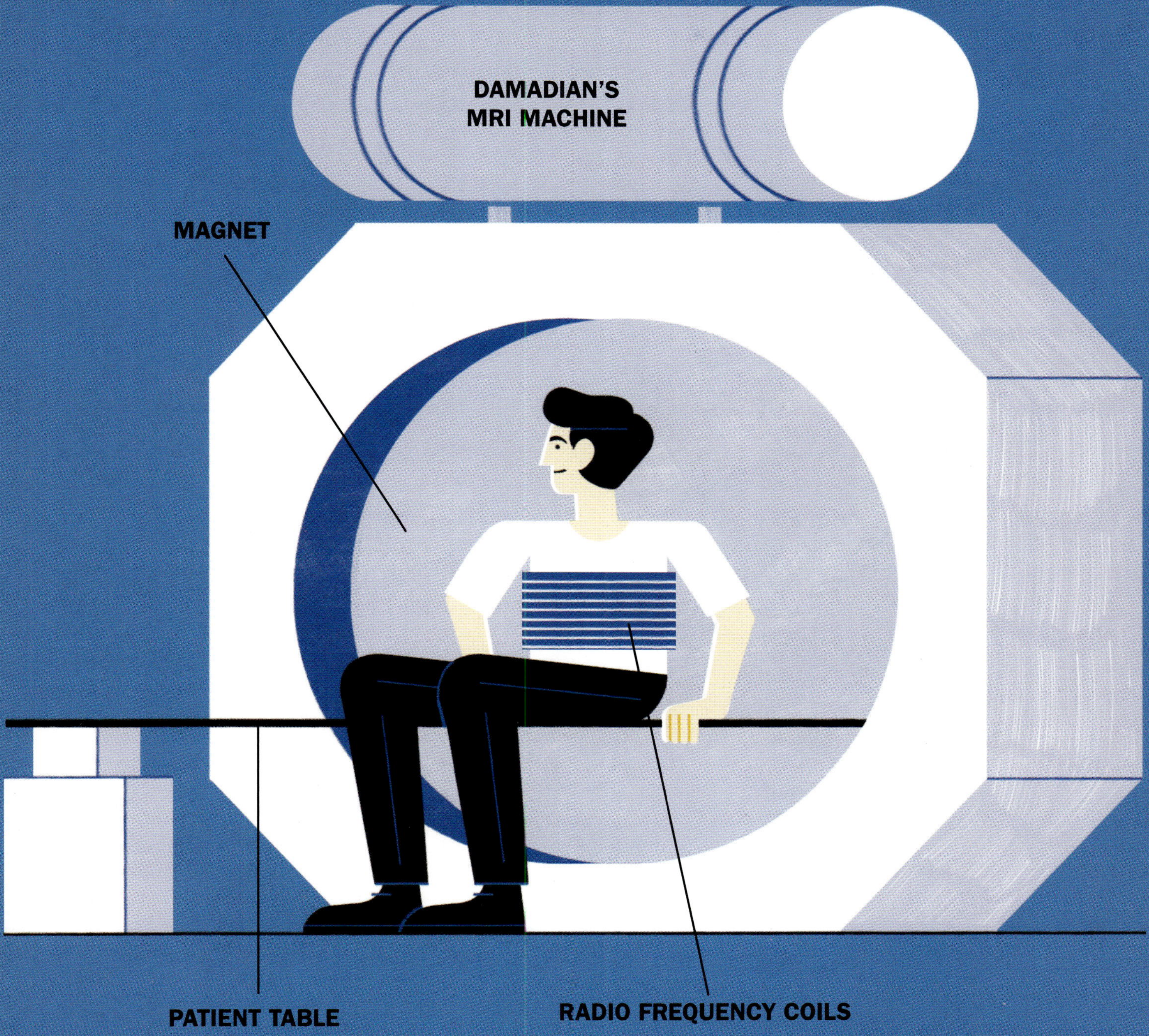

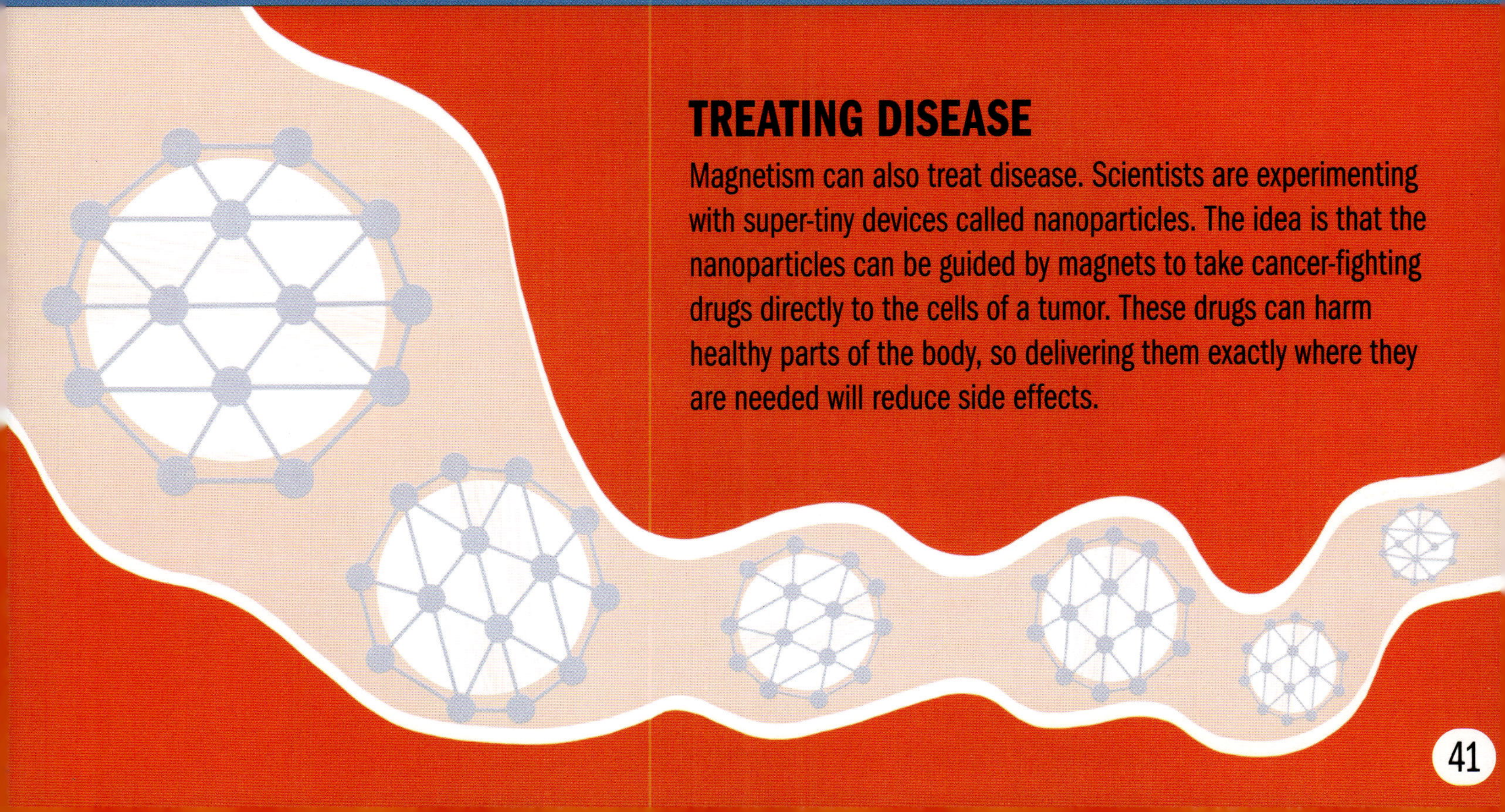

TREATING DISEASE

Magnetism can also treat disease. Scientists are experimenting with super-tiny devices called nanoparticles. The idea is that the nanoparticles can be guided by magnets to take cancer-fighting drugs directly to the cells of a tumor. These drugs can harm healthy parts of the body, so delivering them exactly where they are needed will reduce side effects.

BUILD MAGNET GAMES

YOU WILL NEED:
- 2 disc magnets
- tape
- cardstock pieces 30–48 from the kit

MAZE GAME

You've seen how useful magnets can be, but they can be fun, too! Can magnetism lead a robot through a maze?

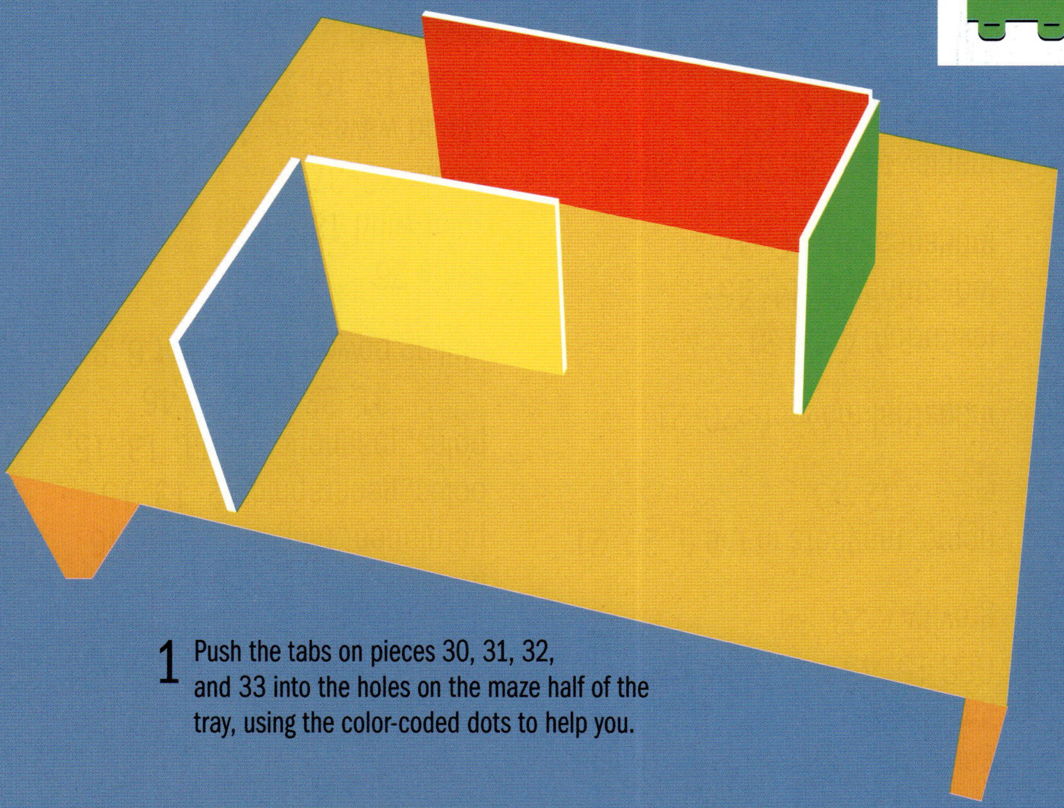

1 Push the tabs on pieces 30, 31, 32, and 33 into the holes on the maze half of the tray, using the color-coded dots to help you.

2 Use pieces 34 to 42 to secure each tab underneath the tray.

3 Stick piece 44 underneath piece 43 with tape. Place the tab on the robot (piece 45) into the slot on piece 43, so that your robot stands freely. Slot piece 46 into the robot, to secure in place.

HINT
If your magnet wand isn't working, try flipping the magnet over. Remember, opposite poles attract.

4 Use a piece of sticky tape to hold a magnet underneath your robot in the space of piece 44.

5 Stick piece 47 underneath piece 48. Place the other disc magnet into the hole on piece 48 and use another piece of tape to hold it in place.

6 Stand your robot on the start arrow on the maze. Hold the handle of your magnet wand and place the other end underneath the maze. Move your wand around to move the robot through the maze, to reach the end flag.

GO FISHING

You may not be able to see which of these fish have a paperclip on the underside. But with a magnet, finding them is a breeze!

YOU WILL NEED:
- 1 disc magnet
- 5 paperclips
- tape
- string
- cardstock pieces 3–8 and 49–58 from the kit

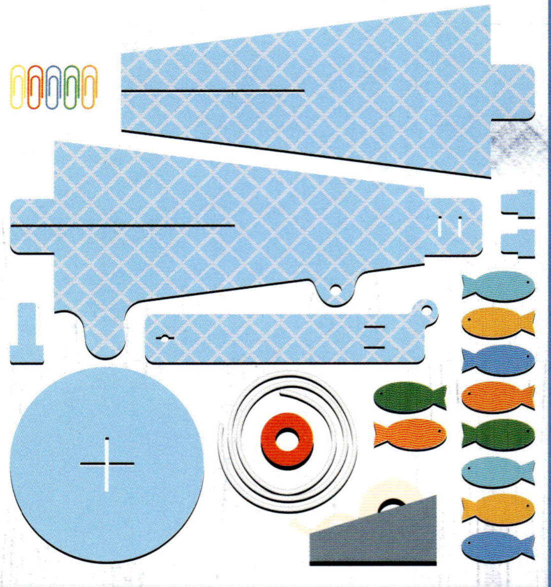

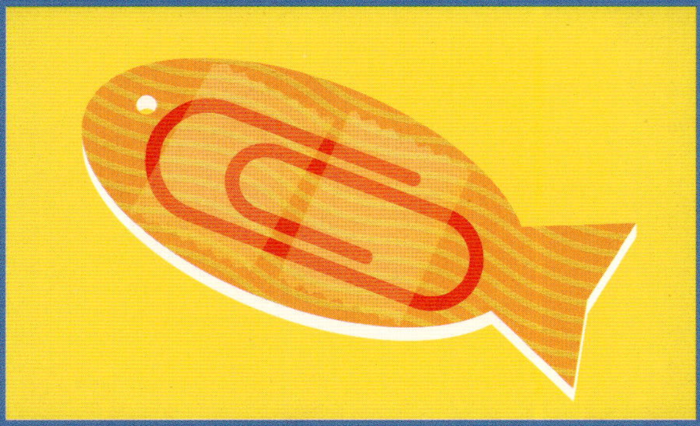

1 Take pieces 49–54 and attach a paperclip to five of the fish.

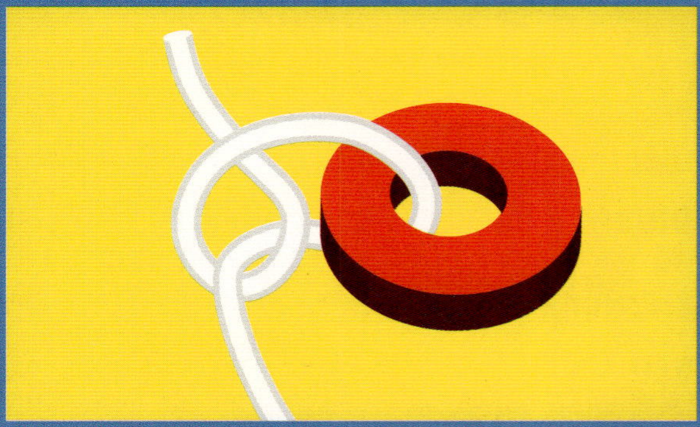

2 Tie one end of the string to one of the disc magnets.

3 Lay some of the fish, paperclip side down, near the hole underneath the crane arm. See page 22 for how to build the crane arm.

4 Pass the other end of the string through the holes on the crane (see page 23).

5 Use the crane to try to pick up as many fish with the paperclips on as you can in a certain time. Challenge your friends to beat your score.

HINT

Magnetic field lines can easily travel through the thin cardstock that the fish are made from. They don't block the magnet on the crane from attracting the metal paperclips.

GLOSSARY

Atmosphere the blanket of gases that surrounds the earth

Aurora a shimmering light in the sky caused by high-energy particles from the sun interacting with Earth's magnetic field

Carbon dioxide a gas found naturally in the atmosphere, which is also produced when we burn fossil fuels

Compass a tool with a magnet that points north, which helps people find their way

Core the very center of Earth, made of extremely hot iron and nickel

Counterfeit fake; not genuine

Data information that is created, stored, and used on a computer in the form of electrical signals

Domain a small region of magnetism within a larger piece of magnetic metal

Electric motor a device that uses magnetism to convert electrical energy to movement energy

Electromagnet a temporary magnet formed by passing an electric current through a coil

Force an influence, such as a push or a pull, that can produce an effect on an object

Friction the resistance caused when one surface or object moves against another

Gauss a unit for measuring magnetism; 10,000 gauss is equal to 1 tesla

Generator a device that uses magnetism to convert movement energy to electrical energy

GPS (short for global positioning system) a network of satellites in orbit around Earth that can be used to pinpoint the location of a receiver on the surface

Lodestone a piece of magnetite, a naturally occurring magnet

GLOSSARY

Maglev train a train that uses powerful magnets to levitate above the tracks in order to reduce friction

Magnetic field the region surrounding a magnet, where its force can be felt

MRI (short for magnetic resonance imaging) a tool for using magnetism to make images of the inside of the human body

Nanoparticles extremely small particles of matter

Permanent magnet a material that stays magnetic over a long period of time

Pole one of the two ends of a magnet, where magnetic forces are strongest

Radiation the emission of energy as waves or particles; some types of radiation can be harmful

Repel to push away

Satellite an uncrewed spacecraft that stays in orbit around a body, such as Earth

Solar wind a stream of tiny charged particles that spews out from the sun in all directions

Sound wave a wave that carries sound energy through air or another substance

Temporary magnet a substance that only acts as a magnet when it is near a permanent magnet or has an electric current running through it

Tesla a unit for measuring magnetism; 1 tesla is equal to 10,000 gauss

Turbine a machine that spins when a flow of air, water, or another substance flows over its blades

Vibrate to move very quickly back and forth

INDEX